Feb 25, 2015

To Tom,

Best wishes

[signature]

THE
POWERHOUSE

Also by Steve LeVine

Putin's Labyrinth
The Oil and the Glory

THE
POWERHOUSE

INSIDE THE INVENTION OF
A BATTERY TO SAVE THE WORLD

STEVE LEVINE

VIKING

VIKING
Published by the Penguin Group
Penguin Group (USA) LLC
375 Hudson Street
New York, New York 10014

USA | Canada | UK | Ireland | Australia | New Zealand | India | South Africa | China
penguin.com
A Penguin Random House Company

First published by Viking Penguin, a member of Penguin Group (USA) LLC, 2015

ISBN 978-0-670-02584-8

Printed in the United States of America
1 3 5 7 9 10 8 6 4 2

For Avery LeVine and Muratbek Nurlybayev

Contents

PART I. THE STAKES

PART II. FOREIGNERS IN THE LAB

PART III. RECKONING

CONTENTS

PART I

THE STAKES

Jeff Chamberlain's War

Wan Gang worried Jeff Chamberlain. Before returning home to Beijing, Wan, China's minister of science, had asked to visit two places—Argonne National Laboratory, a secure federal research center outside Chicago, and a plant near Detroit where General Motors was testing the Volt, the first new electric car of its type in the world. Jabbing his finger into a book again and again, Chamberlain said that Wan was no mere sightseer. He had a mission, which was to stalk Chamberlain's team of geniuses, the scientists he managed in the Battery Department at Argonne. They had invented the breakthrough lithium-ion battery technology behind the Volt, and Wan, Chamberlain was certain, hoped to appropriate Argonne's work. But Chamberlain was not going to let him. A war was on, he said—a battery war. And he was right.

Wan turned up at Argonne in the summer of 2010, animated and unfailingly polite, with gentle eyes and the look of his fifty-eight years. A senior Department of Energy official climbed onto a bus alongside him and his retinue for a tour of the laboratory, and Wan posed a fusillade of questions while offering his own observations. "We are experimenting with the creation of hydrogen fuel from the gas created by waste," he said. "It costs half the price of gasoline." Such talk charmed the battery guys. He was himself a materials scientist, with his own record of advances, speaking openly with equals. It helped that Wan did not explicitly mention

nickel manganese cobalt, or NMC, the compound at the core of
the Argonne invention contained in the Volt. In addition, he had
quite a personal story. Growing up in poverty in the countryside
surrounding Shanghai, Wan recalled going hungry and navigating
fields in a tractor, the only motor vehicle he ever drove. From
there, he worked a series of research jobs and won his first break—
admission to a Ph.D. program at the Clausthal University of Technology in Germany. After he graduated, Audi hired him as an
engineer and he rose to be design manager in the automaker's
Stuttgart-based electric car unit, an exceedingly prestigious position. In all, he had been working at Audi for eleven years when,
one day, his former academic mentor at Shanghai's Tongji University visited the plant. He suggested that Wan transform his own
country, and not Germany, into an electric-car powerhouse. Wan
returned to China, where another break came: President Hu Jin-
tao requested that Wan formulate a policy on electric vehicles and
make China the world's number-one producer of them. He elevated him as the country's first non–Communist Party minister
since the 1950s. Now it was Wan's job to execute Hu's will. The
prevailing view abroad was that, it being China, Wan would succeed. Which brought the Americans back around to their original
angst after having warmed to him.

The evening before his visit to Argonne, Wan was munching
shrimp hors d'oeuvres at a reception on the terrace of the Kennedy
Center in Washington, D.C., when an American recognized and
approached him. Wan seemed to have been waiting for just this
chance conversation. He took a last bite and passed the tail to an
aide. "Why don't we sit over there," he said, gesturing to the café.
They exchanged talk on personal topics, and when they turned to
cars Wan said he agreed that a race was under way among industrialized nations. All of them were determined to create a great
new battery that in turn would propagate the large-scale manufacture of electric vehicles. They were merely using different methods
to get there. Wan was too genteel to predict outright that China

would win but cited markers that would signal progress. "The big thing is getting the first one percent of the market," he said, meaning 150,000 electric cars on China's roads. "That will prove the technology. From there, it won't be that hard to reach ten percent of the market three or four years after that." His initial goal was the sale of 500,000 cars, about the same aim as Barack Obama had established for the United States, and one million by 2015. It was a lot of cars. But the numbers also reflected bravado. Both countries inflated their numbers to impress and psych out rivals.

The next morning at Argonne, Wan and his hosts filed into a conference room. A senior American scientist named Al Sattelberger led off the presentation. He flashed slides on two large screens. Wan interrupted.

"You have made remarkable achievements here," he said. "So today I have many questions for you."
"That's why I'm sweating," said Sattelberger.

The room erupted in laughter. It was mostly the Americans, who *were* sweating. Argonne possessed formidable intellectual firepower and inventions, such as the American patent for its NMC breakthrough. It achieved three grand aims—allowing the Volt to travel forty miles on a single charge, to accelerate rapidly, and to do both without bursting into flames. But despite the recent accomplishment, the United States trailed far behind its rivals. After more than a decade of manufacturing, Japan and South Korea controlled two thirds of the market for consumer batteries such as AAs, AAAs, and the lithium-ion technology used in smart phones. That gave them preeminence on the proving ground where new technologies were validated or broken: the factory floor. Most winning inventions became so when the kinks were worked out through trial and error with actual consumers—what the Japanese and South Koreans had done—and otherwise might be destined for oblivion. Now, the Chinese had adopted the principle and issued an edict requiring some two dozen companies to market models within two or three years. That had led Chinese manufacturers like BYD, Chery, and Geely to introduce experimental electric vehicles. None of China's rivals, the United States included, could simply decree the manufacture of one million electric cars with the confidence that they would actually appear. China's leaders had accomplished innumerable such feats. They terrified Jeff Chamberlain.

Why Argonne Let Wan In

One might fairly ask why Wan was allowed to visit Argonne. The perverse rationale was that the United States *was* so far behind. The Americans resembled the Japanese in the 1970s and the Chinese in the 1990s—they were very much at the bottom of a learning curve others had scaled before. Given that reality, the shrewdest path was to humbly work with the best in the world, glean what you could in visits such as Wan's, then trust in intellectual brawn to push through to victory.

The global meltdown of 2008 and 2009 had put a scare into Americans, who were determined to build a fresh economy on a foundation of substance and not financial, real estate, or dot-com bubbles. Europeans were similarly fearful and determined not to be left out of such a new frontier. Asia's export-propelled economies knew they, too, had to find another way. History told Wan Gang that global financial crises breed the type of fundamental technological discoveries that move economies. He observed before him the makings of just such a breakthrough in energy technology. Like the Americans and Europeans, Wan said that powerful, affordable batteries and the cars they propelled were bound to initiate the next great economic boom. Batteries were an underappreciated technology—they were already enabling the revolution in electronic devices, he said, and now were on the cusp of much more.

Others focused on how a transformed battery could shake up

geopolitics. An electric age would puncture the demand for oil and thus rattle petroleum powers such as Russia's Vladimir Putin, Saudi Arabia's ruling family, and the Organization of the Petroleum Exporting Countries as a whole, stripped of tens of billions of dollars in income. China could put its population in electric cars, shun gasoline propulsion, and clean up its air. Generally speaking, the world might spend less on oil and worry less about climate change.

The numbers behind all this maneuvering were large. Forecasts of the annual market for advanced batteries in 2020 were about $25 billion, half the 2012 gross revenue of Google.[1] That sum would double in the likely event that oil prices settled near or in the triple digits per barrel and drove more motorists away from gasoline propulsion. Battery-enabled hybrid and electric vehicles would command sales of $78 billion by 2020.[2] If large-scale batteries could economically store electricity made by windmills and solar cells, that would be tens of billions more in annual sales.

Yet those figures accounted only for the *current* decade. The general thinking was that, after 2020, the new industries would be even more gargantuan, on the scale of today's ExxonMobil, General Electric, and Toyota, the kind of rare, high-value enterprises capable of firing up an entire future economy. By 2030, advanced battery companies would swell into a $100 billion-a-year industry and the electric car business into several $100 billion-a-year behemoth corporations.[3]

When you sought justification for this enthusiasm, you heard a mainstream assumption that hybrid and pure electric vehicles would make up 13 to 15 percent of all cars produced around the world by 2020; a decade or two later, they would reach about 50 percent.[4] These estimates did not seem unreasonable when you considered the twenty- and thirty-year-long sales trajectories of previous consumer juggernauts like laptops and cellular phones.

Regardless of the care with which they were calculated, the sums were mischievous—no one could accurately project the market for products that did not yet exist. But the leaders of most of the world's industrialized countries—Japan and South Korea,

Brazil, Finland, France, Germany, Israel, Malaysia, Russia, Singapore, South Africa, and the United Kingdom, not to mention the United States and China—decided it was a race, and so it was. In the words of a French government minister, it was a "battle of the electric car."[5]

Because of its record for executing goals at large scale, China loomed over the contest. Yet the Argonne guys felt comfort in that China was not there yet. For one thing, it was still cranking out second-rate technology. Japanese companies, with their two-decade manufacturing lead, conversely enjoyed a commanding 43 percent of the global market for lithium-ion batteries. South Korea held another 23 percent. As for the United States, some people counted it out, but not many. Because lithium-ion had theoretical room for more or less double its current performance, and the United States had both serious scientists and a large market, it still had considerable room to try.

A senior Argonne scientist said that when Wan toured the lab, the undercurrent was, "How can we benefit from this visit?" This created a double intelligence game. Lab managers focused deliberately on the snippets of conversation in which Wan might tip his hand. Yet they could push the boundaries of politesse. During his turn at the podium, for instance, Chamberlain mentioned a clutch of German, Japanese, and South Korean companies—BASF, Panasonic, Samsung, LG Chemical—that were reconfiguring their batteries with the NMC. They were seeking twice the energy of the lithium-iron-phosphate compound favored by Chinese battery makers. Chamberlain was sure that Wan already knew this, making the remark simple candor. If Wan perceived a dig at Chinese strategy, his expression did not betray it. Having gained privileged access to the lab, he rather seemed extraordinarily attentive as he listened to Argonne's history and inspected some of its crown jewels.

A Good Place to Do Science

Although venture capitalists and other titans of Silicon Valley could belittle government-run science, they spoke differently about the Department of Energy's seventeen national laboratories. Argonne commanded particular respect because of its past. It went back to 1942, when Nobel laureate Enrico Fermi traveled to Chicago as the Manhattan Project was getting under way. Fermi set up a makeshift laboratory underneath the Stagg Field football stadium at the University of Chicago and called it the "Met Lab," for Metallurgical Laboratory. Obsessed with secrecy, he and his collaborators kept even their wives uninformed of the big breakthrough—Fermi's creation of the world's first self-sustained nuclear chain reaction, which began the nuclear age. Their sole disclosure went in code to the project leader: "The Italian navigator has just landed in the New World."

"Were the natives friendly?" came the planned reply.

"Everyone landed safe and happy."[1]

Fermi then moved on to Los Alamos to help build the world's first atomic bombs and the Met Lab went on without him.

Eighty-nine-year-old Dieter Gruen had worked at Argonne for six decades, since almost the beginning of the Stagg Field days. "That's Glenn Seaborg," he said in his office, pointing to a framed photo of the cocreator of plutonium. Gruen was smallish and wore a silk, herringbone blazer. When he was fourteen, Gruen and his older brother fled Nazi Germany and made it to the United States. Gruen ended up attending high school in Little Rock, Arkansas, then Northwestern University, where he studied physics. In 1944, he turned up at Stagg Field with a bachelor's degree. He was twenty-one. World War II was at a critical stage—D-Day had just happened—and young people like him were in high demand by Manhattan Project managers. He was dispatched immediately to Oak Ridge, Tennessee, to help produce sufficient uranium-235 for shipment to the bomb makers at Los Alamos, an effort that was behind schedule.

Gruen found some thirty thousand people already at Oak Ridge. The town had been built practically overnight just for them. There was a sea of mud. Construction was everywhere. Gruen slept in a barracks known as West Village 54. Enormous machines called calutrons had been built to produce uranium-235. Oak Ridge had been chosen because it was near powerful Norris Dam, the first big project of FDR's Tennessee Valley Authority, which could provide the immense volume of electricity that the calutrons required.

So it went for eighteen months, until the war ended with the atomic bombing of Hiroshima and Nagasaki. The work at Oak Ridge wound down. Gruen returned to Stagg Field while beginning graduate studies at the University of Chicago—the Met had been named the country's first national laboratory and had plenty for him to do. There was so much activity, in fact, that the Met felt cramped for space. Lab scouts began to hunt for a new home. They settled on a place called Tulgey Wood, a two-hundred-acre spit of farmland twenty-four miles southwest of the city along Route 66.

In 1936, Erwin O. Freund, a sausage titan who invented the skinless hot dog, named Tulgey Wood, his new estate, after the forest in *Alice in Wonderland*. Freund was extravagant and eccentric. He placed small, painted carvings of Tweedledee, Tweedledum, and other Lewis Carroll characters along bark trails through the property. He kept two pet chimps plus sheep and peacocks, raised championship boxers in an air-conditioned kennel, and dug limestone-lined lakes for boating in summer and ice-skating in winter. When a clothier friend gave Freund seven fallow deer—a species called *Dama dama*, which are born tan but in adulthood turn completely white—he cared for them, too.

Freund put up a fight when he learned that the Met scouts had settled on Tulgey Wood as the lab's new home. He decided to employ "every means at my command for as long as necessary to prevent its being seized from me."[2] The government's intent was to buy, not seize, the property, yet Freund battled to keep his estate. The dispute went on for a year, when, in 1947, Freund died of a sudden heart attack, allowing the federal acquisition to proceed.

The boxers were easy to move, but the deer would have to go to game parks. Some simply could not be captured and were left behind to wander. Over time, the scientists noticed the herd growing back, "glimpsed along the tree line through a morning fog, found on a knoll during an evening rain, or spotted in headlights near a road at night."[3] They became an enduring remnant of Erwin Freund's grand project.

But what to officially call the Met now that it occupied a new place? Someone suggested Fermi Lab, but since such dedications ordinarily honored the deceased and the scientist was still living, the name of a local town was selected—Argonne.

The government purchased additional surrounding farmland. Argonne now covered 4,100 acres. To fill it in, workers planted about a million pine seedlings, which thrived and created a massive home for the growing deer herd. Argonne still looked like a military base, dotted with Quonset huts erected as offices. In the 1950s, red brick structures were added. They were given numbers instead of names. Building 205 was finished in 1951. The two-story structure would become home to Argonne's Battery Department.

Many of Argonne's first scientists commuted from Chicago aboard a shuttle bus for thirty-five cents' fare. The lab provided the service because nearly all its staff lived in the city. Some called Hyde Park "Little Argonne" because of the number of residents employed by the lab. The ride took ninety minutes and passed streets dense with factories, warehouses, and rail yards before giving way to an expanse of farmland. It might seem long, but the driver, an amateur ventriloquist, entertained as he went. In one trick, he would startle boarding passengers with a voice that suggested someone shouting from behind to get on. But gradually the shuttle was discontinued as the scientists gave up the city and sought homes in nascent suburbs such as Aurora, Naperville, and Downers Grove. These communities, with roots stretching back to the 1830s, often resisted newcomers, and Argonne would have to vouch for their character before they could move in. Yet eventually most were accepted and some even found themselves embraced. Among the latter was Stephen Lawroski, head of the Chemicals Technology Division, whom tony Naperville dubbed "the Professor" and honored with a regular invitation to a daily breakfast club of local dignitaries at a downtown drugstore.

Dieter Gruen was awarded his doctorate in 1951. Graduates with his background had many choices. Fundamental research was under way across American industry. He interviewed at AT&T's Bell Laboratories and heard of positions at General Electric, Ford, and General Motors. Universities, too, were hiring professors and basic researchers. But Gruen remained drawn to Argonne, where he was already known and still proud to work. Argonne was already one of the world's premier research facilities. Experimentalists enjoyed a free flow of funding from Washington and tremendous liberty to research what interested them. Gruen accepted badge number 1989 and an office in Building 205.

At first, Gruen was assigned to a team building a nuclear submarine under the direction of Captain Hyman Rickover. His task was to figure out how to eliminate hafnium from zirconium, needed in combination with uranium to fuel the subs. The regime was strict. Virtually everything was top secret given that Argonne's primary function was to create sensitive nuclear technology. Gruen felt the danger. Scientists wore special yellow shoes and provided regular urine samples, both precautions against radiation contamination. An eight-foot fence surrounded the building, accessible only through a guard post. Every office contained a red wastepaper basket with bold all-capital lettering: BURN. They were for the classified papers that were no longer wanted. You weren't supposed to incinerate such documents yourself—the label was aimed at the disposal staff. But at least once, a scientist took the designation literally, setting his wastepaper basket afire and sending smoke into the hallway.

A couple of hundred people were already working in Building 205. Most of them were in their twenties and thirties, a mix of men and women, the latter mostly secretaries, and many were single. At lunch, the men bet over rounds of pinochle in the basement and through the day frequented coffee groups organized along every corridor. On weekends, the scientists visited one another's homes and numerous couples ultimately married. But generally speaking, Argonne seemed organized for the work conducted there without regard for the conditions under which it was carried out. Only rooms that absolutely required air conditioning were equipped with it, which meant that in the humid summer, moisture collected on overhead water lines and dripped onto the scientists. Some of them draped their equipment with protective plastic but they themselves still often got wet. At departmental meetings, overheated researchers regularly fell asleep.

Gruen didn't find it at all like Oak Ridge—the intensity was not there. After all, the war had ended. If you ignored the dangerous and classified projects under way, the lab seemed ordinary. Scientists worked from nine to five. In 1956, Gruen and his wife moved to Downers Grove, which had become another Little Argonne. "We didn't think anybody lived in Downers Grove except people who worked at Argonne," one of their children remarked.

Yet Gruen also noticed the envy of university friends. He had the use of rare and advanced equipment. If you were a "hotshot," which he was—he was his team's youngest senior scientist and assigned his own research group—you were smart to be at Argonne.

4

"Discouraged Weariness in the Eyes"

A t times, the ferment of the 1960s seemed aimed at Argonne. Seeing their nuclear research stigmatized and budgets reduced, some people thought that Argonne's existence was threatened. After a while, the lab director noticed "discouraged weariness in the eyes" of the scientists. Recalling his own time in Exxon's research lab a few years back, the director reckoned that much of the gloom sprang not from the national politics but Argonne's atmosphere—scientists were likelier to produce first-rate work if they were surrounded by first-rate facilities. He asked his wife to help. Before long, she had workers retiling and painting Building 205. They added lights in the public areas and gussied up the hitherto pale green offices with pinks, golds, and blues. The overall effect was a softer ambience, "a brand new building," especially with the finishing touch of a jazz and blues concert series.

One researcher carried a loaded derringer into the lab, explaining that he attended classes in a dodgy neighborhood and needed the protection. He was fired when the pistol discharged as he changed clothes, wounding him. "No further gunplay in the locker room," the division director said. At the annual Turkey Raffle in the basement auditorium, Sandy Preto, a lab researcher who moonlighted as a belly dancer at a nearby club, surprised colleagues with a performance.[1]

Throughout, the lab's hazards remained unignorable. One day, a new scientist named Paul Nelson assisted a senior researcher who was heating and freezing molten zinc mixed with a few tenths of a gram of plutonium. For protection, they wore gas masks, but the concoction accidentally spilled and burned straight through some hot stainless steel. Nelson "thought about my children and decided it was time to leave." Colleagues subjected him to good-natured ribbing for fleeing a harmless bit of combustion. They were somewhat less casual a few years later when an experiment with uranium and plutonium oxide blew out the glass panels of a working lab, created a bulge in the concrete walls, and scattered radioactivity.[2] Researchers had accidentally installed the safety meter backward, leading to a buildup of hydrogen and oxygen. Cleaning crews removed the contamination while the researchers sat out some time on medical watch.

Some things went unchanged—gazing from his window one day, Nelson counted eighty-three white deer—but Argonne was aging. In the 1970s, a former senior manager remarked that the lab "isn't exactly the Club Med type of atmosphere that one would expect to engender romantic relationships."[3] When high emotions did arise, they seemed to pit the various arms of science against one another. The engineers called the chemists "pharmacists," who assailed the former as "pipefitters." The physicists had a similarly low opinion of materials scientists. But the physicists cast themselves favorably as "part of the big science world [that] thought big." Unlike the energy storage scientists, who insisted on going home for dinner at six, the physicists frequently worked around the clock, through weekends and on holidays if necessary, to repair, say, a failed particle accelerator.[4]

There was truth in what the physicists said—Argonne's battery guys by and large were not the type who stuck out.

That was new, because for much of the eighteenth and nineteenth centuries, batteries and the electricity held within them were treated as an almost unfathomable force by poets, philosophers, and scientists. Those who had unleashed the epoch were accorded tremendous deference. Alessandro Volta invented the first battery and thus launched the electric age in 1799. It was a feat rooted in a debate with fellow Italian Luigi Galvani, who claimed that frogs possessed an internal store of electricity. Volta theorized that the electricity observed by Galvani originated in metals used as part of the experiment, rather than in the frogs themselves. Volta created his battery while carrying out experiments to disprove Galvani. Benjamin Franklin, a contemporary, had already coined the word to describe a rudimentary electric device he built out of glass panes, lead plates, and wires. But Franklin's was a battery in name only, while Volta's was a true electric storage unit. After Volta's brainchild, scientists kept hooking up batteries to corpses to see if they could be coaxed back to life. Many wondered whether electricity could cure cancer or if it was the source of life itself. What if souls were electric impulses?

To make a battery, you start with two components called electrodes. One is negatively charged, and is called the anode. The other, positively charged electrode is called the cathode. When the battery produces electricity—when it discharges—positively charged lithium atoms, known as ions, shuttle from the negative to the positive electrode (thus giving the battery its name, lithium-ion). But to get there, the ions need a facilitator—something through which to travel—and that is a substance called electrolyte. If you can reverse the process—if you can force the ions now to shuttle back to the negative electrode—you recharge the battery. When you do that again and again, shuttling the ions back and forth between the electrodes, you have what is called a rechargeable battery. But that is a quality that only certain batteries possess.

The battery's very simplicity—its remarkably small number of parts—has both helped and hindered the efforts of scientists to improve on Volta's creation. They had only the cathode, the anode, and the electrolyte to think about, and, to fashion them, a lot of potentially suitable elements on the entire periodic table. Yet this went both ways—there was no way to bypass those three parts and, as it soon became apparent, only so many of the elements that were truly attractive in a battery. In 1859, a French physicist named Gaston Planté invented the rechargeable lead-acid battery. Planté's battery used a cathode made of lead oxide and an anode of electron-heavy metallic lead. When his battery discharged electricity, the electrodes reacted with a sulfuric acid electrolyte, creating lead sulfate and producing electric current. But Planté's structure went back to the very beginning—it was Volta's pile, merely turned on its side, with plates stacked next to rather than atop one another. The Energizer, commercialized in 1980, was a remarkably close descendant of Planté's invention. In more than a century, the science hadn't changed.

In the early part of the twentieth century, electric cars powered by lead-acid batteries seemed superior to rivals featuring the gasoline-powered internal combustion engine. But a series of inventions, including the electric starter (which eclipsed the awkward rotary hand crank), finally gave the advantage to the internal combustion engine propelled by gasoline and contained explosions rather than a flow of electricity. For four decades, few seemed to think that things should be different.

In 1966, Ford Motor tried to bring back the electric car. It announced a battery that used *liquid* electrodes and a *solid* electrolyte, the opposite of Planté's configuration. It was a new way of thinking, with electrodes—one sulfur and the other sodium—that were light and could store fifteen times more energy than lead-acid in the same space.

There were disadvantages, of course. The Ford battery did not operate at room temperature but at about 300 degrees Celsius. The internal combustion engine operates at an optimal temperature of about 90 degrees Celsius. Driving around with much hotter, explosive molten metals under your hood was risky. Realistically speaking, that would confine the battery's practical use to stationary storage, such as at electric power stations. Yet at first, both Ford and the public disregarded prudence. With its promise of clean-operating electric cars, Ford captured the imagination of a 1960s population suddenly conscious of the smog engulfing its cities.

Popular Science described an initial stage at which electric Fords using lead-acid batteries could travel forty miles at a top speed of forty miles an hour. As the new sulfur-sodium batteries came into use, cars would travel two hundred miles at highway speeds, Ford claimed. You would recharge for an hour, and then drive another two hundred miles. A pair of rival reporters who were briefed along with the *Popular Science* man were less impressed—despite Ford's claims, one remarked within earshot of the *Popular Science* man that electrics would "never" be ready for use.

The *Popular Science* writer went on:

> They walked out to their cars, started, and drove away, leaving two trains of unburned hydrocarbons, carbon monoxide, and other pollution to add to the growing murkiness of the Detroit atmosphere. [The other reporter's remark] was a good crack. But it was wrong. When a development is needed badly enough, it comes. Without some drastic change, American cities will eventually become uninhabitable. The electric automobile can stop the trend toward poisoned air. Its details are yet to be decided. But it will come. And it won't be long.[5]

For a few years, the excitement around Ford's breakthrough resembled the commercially inventive nineteenth century all over again. Around the world, researchers sought to emulate and, if they could, best Ford. As it had been on nuclear energy, Argonne sought to be the arbiter of the new age. In the late 1960s, an aggressive electrochemist named Elton Cairns became head of a new Argonne research unit—a Battery Department. Cairns initiated a comprehensive study of high-temperature batteries like Ford's. Someone suggested a hybrid electric bus assisted by a methane-propelled phosphoric acid fuel cell, and it was examined as well. Welcoming suggestions, the lab director insisted only that any invention be aimed at rapid introduction to the market. To be sure that would happen, he invited companies to embed scientists at Argonne for periods of a few months to a year, and many did so.

John Goodenough, a scientist at the Massachusetts Institute of Technology, said that everything suddenly changed. Batteries were no longer boring. Goodenough attributed the frenzy to a combination of the 1973 Arab oil embargo, a general belief that the world was running out of petroleum, and rousing scientific advances on both sides of the Atlantic. Pivoting off the Ford work, a young British chemist named Stan Whittingham, working as a postdoctoral assistant at Stanford University, discovered that he could electrochemically shuttle lithium atoms from one electrode to the other at room temperature with inordinate damage to neither. To explain this action, which created rechargeability, Whittingham borrowed the term *intercalation* from chemistry, and it stuck. Exxon, the oil giant, wishing to compete with Bell Labs—"to be perceived as *the* lab of the energy business"—offered to hire Whittingham at a significant salary.[6] He accepted and set out to make a battery from his findings.

Whittingham was drawn to lithium, silvery white and malleable, because it is the lightest metal on the periodic table. But it reacts with air and, in certain circumstances, catches fire. Scientists therefore handle pure lithium metal only in a laboratory setting in which all moisture has been removed from the air. Whittingham could make lithium metal practical only if he could combine it with another metal into an alloy—which is what he did, coupling it with aluminum to create a small and powerful anode. In 1977, Exxon released Whittingham's device as a promotional product, a coin-size battery that fit in the back of a solar watch. It was the first rechargeable lithium battery. But when Whittingham tried to make them larger, his batteries kept igniting in the Exxon lab. Despite the presence of aluminum, the lithium metal was still too reactive.

Then Goodenough, the MIT scientist, proceeded to outdo all that Ford, Argonne, and Whittingham had accomplished. By the time he was finished, he would either himself produce, or be part of the invention of, almost every major advance in modern batteries.

5

Professor Goodenough

J ohn Goodenough grew up in a sprawling home near New Haven, Connecticut, where his father, Erwin, was a scholar on
the history of religion at Yale. His parents' relationship "was a
disaster," he said, friction that extended into aloofness toward
their children; Goodenough and his mother, Helen, especially
"never bonded." When he was twelve, John and his older brother,
Walt, were sent to board on scholarships at Groton and he rarely
heard from his parents again. John's mother wrote just once as he
grew to adulthood. In a slender, self-published autobiography,
Goodenough cited many influences: siblings, a dog named Mack,
a family maid, long-ago neighbors. But in this regard he conspicu-
ously ignored his parents and never mentioned them by name.
Theirs was a solely biological place in his life.

Goodenough's boyhood did not suggest the warm, amusing,
and self-assured adult to come. Suffering from dyslexia at a time
when it was poorly understood and went untreated, Goodenough
could not read at Groton, understand his lessons, or keep up in the
chapel. Instead, he occupied himself in explorations of the woods,
its animals and plants. Somehow everything came together. He
went on to thrive at Yale, from which he graduated summa cum
laude in mathematics, then by happenstance fell into science: after
World War II, Goodenough, by then a twenty-four-year-old Army
captain posted in the Azores Archipelago off the coast of Portugal,
received a telex ordering him to Washington, D.C.—educators had

stumbled on unspent budget money and advocated using it to send twenty-one returning Army officers through graduate studies in physics and math. Goodenough had taken almost no science as an undergrad but, for reasons obscured by time, a Yale math professor had added his name to the group. So he found himself at the University of Chicago, studying physics under professors Edward Teller, Enrico Fermi, and others. As Goodenough registered for preliminary undergraduate classes, necessary to catch up with the others, a professor remarked, "I don't understand you veterans. Don't you know that anyone who has ever done anything significant in physics has already done it by the time he was your age?"

But it turned out that Goodenough had an intuition for physics. After obtaining his doctorate in 1952, he went to work at MIT's Lincoln Laboratory, which the U.S. Air Force had funded the year before to create the country's first air defense system. His team was told to invent a system of computer memory, a vital component of the envisioned air defense, which was to be called SAGE. At the time, computers comprised enough vacuum tubes to fill "the space of a large dance hall," in Goodenough's words, and had infernally slow memories.[1] Some thought the task impossible because of the physical limits of the ceramic material with which the team was working. Three years later, the lab unveiled an invention that they called "64 x 64 bit magnetic memory," a triumph that, in addition to helping to enable SAGE, became the foundation of later computer memory systems. For Goodenough, more advances followed, including the "Goodenough-Kanamori rules," which became a standard for how metal oxide materials behave at the atomic scale, another building block of future computers.

Politics intruded—a U.S. senator named Mike Mansfield pushed through a law requiring that any research financed by the Air Force have an Air Force application. By now, Goodenough was fixated on finding a scientific answer to the OPEC-led energy crisis, which seemed to be the largest problem facing the country. But he was told to try something else—given the Mansfield law, the subject was the responsibility not of the Air Force but of the national labs.

For Goodenough, it was time to move on. A friend sent word of an opportunity across the Atlantic. Oxford University required a professor to teach and manage its inorganic chemistry lab. Goodenough was surprised to be selected given that he was not a chemist and in fact had completed just two college-level chemistry courses. He was lucky a second time to be chosen for a job for which he was underqualified, on paper.

Goodenough was a tough professor. An early student of his at Oxford recalled a physics course that started with 165 students. After a stern Goodenough lecture, she was one of just 8 to return for the second class.[2] Goodenough was equally exacting in the lab. After MIT, he was on the hunt for big advances in solid state chemistry, a field known for creating the kinds of materials that go commercial. Among the first on his list of targets was Stan Whittingham's recently published breakthrough on the lithium battery.

For six decades, zinc carbon had been the standard battery chemistry for consumer electronics, having eclipsed lead oxide, which was too bulky and heavy for small devices. Whittingham's brainchild was a leap ahead of zinc carbon—powerful and lightweight, it could power portable consumer electronics such as tape recorders. If it worked. But basic physics got in the way. The same electrochemical reactions that enabled lithium batteries also made them want to explode: the voltage would run away with itself, a cell would ignite, and before you knew it the battery was spitting out flames. But you seemed no better off if you played it safe and used other elements—you'd find that they slowly fell apart on repeated charge and discharge.

Goodenough thought he could create a more powerful battery than Whittingham's. Much of invention, he said, involves shifting your mind-set, something many scientists either refused or simply could not do. The Exxon man's battery relied on a sulfide electrode;

Goodenough turned to another family of compounds—metal oxides, a combination of oxygen and a variety of metal elements. In his judgment, oxides could be charged and discharged at a higher voltage than Whittingham's creation, and thus produce more energy. But there was also the matter of getting sufficient lithium to intercalate, the action that created electricity—pulling lithium from a cathode, in this case made of metal oxide, and sending it into a shuttling motion between the electrodes. The more lithium that could be shuttled, the more energy the battery would produce. But it seemed axiomatic that you could not remove all the lithium, because that would leave the cathode virtually hollowed out, and it would fall in on itself. So could any of the oxides manage to hold up under this abuse? And if so, which one, and what was the magic proportion of lithium that could be pulled out?

Goodenough directed two postdoctoral assistants to methodically work their way through structures containing a group of oxides; he asked them to find out at what voltage lithium could be extracted from the oxides, which he expected to be much higher than the 2.2 volts Whittingham was using, and to determine how much lithium could be intercalated in and out of the atomic structure before it collapsed. Their answer was half—about 50 percent of the lithium could be pulled from the cathode at 4 volts before it crumpled, which was plenty for a powerful, rechargeable battery. Of the oxides they tested, the postdocs found that cobalt was the best and most stable for this purpose.

In 1980, four years after Goodenough arrived at Oxford, lithium-cobalt-oxide was a breakthrough even bigger than Ford's sodium-sulfur configuration. It was the first lithium-ion cathode with the capacity to power both compact and relatively large devices, a quality that made it far superior to anything on the market. Goodenough's invention changed what was possible: it enabled the age of modern mobile phones and laptop computers. It also opened a path to the investigation of a potential resurrection of electric vehicles.

Over the years, Goodenough would attract a constellation of bright people to his lab, researchers who often had their best professional years with him. It was not that Goodenough himself did any of the hands-on experimentation—the postdoctoral assistants and researchers he attracted were actually at the bench. Goodenough could be stern, but the atmosphere of big expectation he created drove them to do exceptional work. And he talked them through their projects. One of these researchers was a young South African who arrived in 1981 with a curious idea about gemstones.

6

The Double Marathoner

The Comrades Marathon extends to the South African port of Durban from Pietermaritzburg, twenty-eight miles inland and three thousand feet lower in altitude. The first time that Mike Thackeray ran the race, in 1968, he finished in ten hours and three minutes, just under the eleven-hour cutoff for the slowest participants. Determined to do better, he ran it again. And again. In 1976, entering the race for the fourteenth time, Thackeray took fourth place with a time of 6:32. His discipline had paid off.

Thackeray was the lead inventor of Argonne's NMC technology, a descendant of the lithium-cobalt-oxide cathode pioneered by Goodenough and the formulation that had beguiled Wan Gang. Thackeray's office was situated within the main Battery Department suite, two doors down from his boss Chamberlain. Long halls lined with linoleum and pale green brick walls gave Building 205 a lingering feel of the 1950s. A handwritten sign taped to a coffee brewer requested that drinkers leave behind thirty cents a cup.

Two portraits decorated the walls in Thackeray's office—an 1861 etching of the nineteenth-century physicist Michael Faraday and a sketch of the astronomer William Herschel, who in 1781 discovered Uranus. Thackeray received them as gifts in his youth in South Africa. His mind returned often to his native land, which seemed to speak the most for his soul. Few knew it, he would say, but for a short time almost four decades before, South Africa was one of the great centers of battery thinking.

In Pretoria in the late 1970s, Thackeray, in shaggy, blondish hair and long sideburns, did his Ph.D. under a crystallographer named Johan Coetzer. One day, Coetzer walked into the lab and announced a new project. They were going to "do some stuff in the energy field." The Yom Kippur War between Israel and its Arab neighbors had triggered an energy crisis and the Western world was seeking a way around Middle East oil. Coetzer thought one answer was the advancement of batteries and he told Thackeray that that was where they would focus their work. The effort was challenged from the beginning because of South Africa's system of apartheid, to which the world had responded with economic sanctions. No one outside the country would collaborate with them. To avert international trouble, they had to cloak their work in secrecy and communicate using code words. The smokescreen did not seem to matter much since neither Coetzer nor Thackeray knew anything about energy storage. But their fresh eyes turned out to be advantageous. Approaching the field laterally, "uncontaminated by how other scientists were looking at the world," as Thackeray put it, they found insights into high-temperature batteries, the breakthrough reported by Ford and Stanford. The early result was the Zebra, South Africa's own molten battery. Corporate funding quickly followed, a respectable achievement when you recalled their modest start.

Considering the Zebra, Thackeray thought there still must be a way to do better and at the same time move ahead of John Goodenough's blockbuster advance in 1980. The Zebra and other molten batteries, operating at 300 degrees Celsius, were unsafe, inside a car anyway. As for Goodenough's room-temperature formulation of lithium-cobalt-oxide, it was an improvement but still expensive if you thought of using it in electronic devices.

In physics, there is a structure called *spinel*. These structures have considerable advantages. They are abundant and therefore cheap. They have an appealing three-dimensional structure resembling a crystal. And spinels are inherently stout—sturdier, for instance, than the layered structure of Goodenough's lithium-cobalt-oxide electrode. Goodenough had been instructing his lab assistants to put half the lithium in motion between the cathode and the anode; but Thackeray wondered whether *all* the lithium could be pulled in and out of a spinel cathode. If he could do so without the cathode's collapsing, spinel would be less expensive and potentially much more powerful than the lithium-cobalt-oxide.

The particular spinel that interested Thackeray was iron oxide. Ordinarily, we know iron oxide as rust—it is what happens when you leave your bicycle out in the rain. But for battery scientists, iron oxide is also a spinel, lending it special characteristics. In South Africa, Thackeray had already successfully shuttled lithium in and out of iron oxide working at the same high temperatures as the Ford researchers. He had a hunch that iron oxide might also cooperate at room temperature, which would make it much more practical.

South Africa was disconnected geographically as well as politically. It was almost as far as you could be from the intellectual hubs of the United States and Europe. That being the case, it was almost expected that any self-respecting young South African scientist would spend a year or so abroad. Thackeray decided that he wanted to use his own sabbatical to test his ideas with spinel. And to do so with the leading figure of the day—Goodenough.

Thackeray wrote to Oxford. Goodenough responded immediately: he lacked money to support the younger man, but if that didn't bother Thackeray, he would be pleased to play host. Thackeray, owed his lab-funded obligatory time abroad, required no outside funding. So, at thirty-one, he along with his wife and daughter packed for a fifteen-month postdoctoral assistantship in England.

Thackeray wandered the Oxford campus. As he looked around, he recalled his father's stories of undergraduate study. Generations of Thackerays had attended Cambridge—his father, Andrew David Thackeray, who rose to be a leading astronomer; his grandfather,

Henry St. John Thackeray, a biblical scholar who went on to teach at the university; and of course the novelist William Makepeace Thackeray, a fifth cousin once removed. Both Thackeray and his brother had elected to remain in South Africa for university, and he sensed himself unequal to England's great academic institutions. The intimidation was not just Oxford, but Goodenough himself. Thackeray found the older man's intelligence almost overpowering. He had never heard anything quite like the professor's resonant hoot, an unusual chortle that Goodenough often let fly. By comparison, Thackeray regarded himself as an ordinary "bush chemist from Africa."

He felt out of place for yet another reason—because of his country's medieval political system, he was certain that those he encountered, while saying nothing, must be harboring repulsive thoughts about him and his family. But Goodenough himself plainly did not hold Thackeray or his family personally responsible for the sins of their country. Thackeray would have his own, very different proving ground: the lab.

Thackeray had brought samples of magnetized iron oxide spinel with him from Pretoria. He planned to intercalate lithium in and out of them at room temperature and thus demonstrate that iron oxide spinel could be a powerful new battery material, one commercially superior to the cobalt formulation.

Goodenough dismissed Thackeray's hypothesis. It violated physics—spinels resemble semiprecious gemstones and, structurally speaking, you could not move lithium in and out of a gemstone, the older man reminded Thackeray. Its physical structure, unlike cobalt oxide, would block any such attack.

Thackeray recalled how, despite Goodenough's skepticism, he had already managed the deed at high temperatures back in South Africa. This was an issue of merely lowering the temperature.

"Well, you are welcome to try," Goodenough said. "But you'll want to look around the lab for other stuff to do." Then he left for holiday in India.

Two weeks later, Goodenough returned. "I intercalated the lithium," Thackeray said.

"What?"

The older man took Thackeray into his office and listened as he explained how, using a magnetic stirrer, an automated device for mixing chemicals, he had combined lithium with iron oxide at room temperature. Thackeray observed an immediate encouraging sign—the iron oxide fell away from the stirrer, showing that it had lost its magnetic qualities. That suggested that the spinel had ingested the lithium. Yet this was still not conclusive evidence of intercalation. There had to be more if Thackeray was to state definitively that this was the case. He smeared the concoction onto a glass slide. Now he shot X-rays through it. When you take such pictures you get a spectrum of peaks rather than an image. The trick is to infer the precise structure of the compound from the pattern of the peaks. This skill, the knowledge of X-ray diffraction, is known as crystallography.

Thackeray had shot two X-rays—one prior to the experiment and one after. Comparing them, he noticed "striking differences" in both the position of the peaks and their relative intensity. Something had happened. If Goodenough was right, and the lithium found no entry into the iron oxide structure, the X-ray patterns would be identical. But the peaks had noticeably changed—the lithium *had* intercalated into the iron oxide. Iron oxide spinel could be fashioned into a lithium-ion electrode.

In fact, as Goodenough had postulated, there *wasn't* space for the lithium in the spinel. What Thackeray had shown was that the spinel had an unexpected quality of hospitality—when you moved lithium in, the iron ions shifted around to accommodate it. They made extra space. The spinel experienced a "phase change," absorbing the iron and transforming into a slightly different material resembling rock salt. Like Goodenough's lithium-cobalt-oxide breakthrough a year earlier, Thackeray had conceived of a way to significantly improve on the energy density of zinc carbon batteries. Goodenough was surprised and enthusiastic—Thackeray's idea proved a new principle and, from the standpoint of cost, was potentially better than his own brainchild.

Yet he had not created a practical battery material—a workable cathode—which was the objective. There was a problem, Goodenough said. Examining the data, he detected a blockage in the spinel. The iron oxide wasn't providing a clear path for sufficient lithium to enter and find a home in the structure before being shuttled out in the charge-recharge cycle. A less-cluttered channel was needed if the material was to be truly useful and not a mere novelty.

Perhaps the problem was the *type* of spinel they were using. A different sort—possibly manganese oxide, which he called by its scientific notation, $LiMn_2O_4$—could lift the logjam and allow the lithium into the right places. Goodenough had intimate knowledge of manganese oxide from his MIT days because his team had used it in their computer memory experiments. He suggested that they swap oxides. Manganese spinel, $LiMn_2O_4$, could prove the path to a cheaper cathode.

In the subsequent days, Thackeray, working in the lab library, prepared the manganese spinel experiment. As he did, Bill David, another new researcher under Goodenough, introduced himself.

In a way, David and Thackeray were equals. They were both postdoctoral assistants and had started work around the same day. But David felt like a "young lad" around the South African, who was six years older. Part of that was Thackeray's deceptively unassuming attitude: he told almost no one why he was specifically at Oxford or his accomplishments thus far. Once, David queried him over lunch about a possible jog together. The conversation seemed to go nowhere. Thackeray barely acknowledged any personal interest in running; he said nothing of the Comrades, nor of his status as one of his country's fastest amateurs. For David, once he came to know Thackeray, this reserve lent him powerful mystery.

From a pure physics standpoint, David was under no spell. Like Goodenough, he felt that Thackeray's work was fundamentally counterintuitive—it broke all the rules. He could not dispute the X-ray crystallography—Thackeray was right despite the physics. But there was the blockage cited by Goodenough. Until the obstruction was removed, the experiment could not be called a masterstroke. David thought he could help. He understood atomic-scale crystallography better than Thackeray. David scrutinized the X-ray diffraction of the $LiMn_2O_4$. The intercalation worked perfectly this time, with an open pathway for the lithium. And the spinel had not been torn apart by the foreign material.

Thackeray *was* right.

He was deliriously content. One day, Goodenough was strolling the halls with Thackeray and said, "You know that this could have commercial value, Mike." Though neither man could put a finger on how the invention might be used, Thackeray repeated the remark in a subsequent call to his South African supervisors, who rushed to London. They drew up a patent application—as the inventors, Thackeray's name was listed first, followed by Goodenough as supervising investigator. The owner of the patent would be the South African Inventions Development Corporation, the intellectual property (IP) arm of the government lab in Pretoria where Thackeray was on staff.

Later, there would be dueling personal accounts about who was more responsible for the spinel breakthrough—Thackeray or Goodenough. The older man would claim credit, suggesting that Thackeray merely followed his instructions. Thackeray would retort that he himself had arrived in Oxford with the spinel samples; he had the big idea. But the warm memories made them respectful friends and, ever the diplomat, Goodenough finally summed it up best: "I don't think he would have done it by himself, and I wouldn't have done it without him."

David said that successful science "is about people and it is about ideas." It is about aspiration. Scientists in places like Oxford had such surpassing ambition to reach the top of their field that success seemed the natural result. That was the fun and visceral excitement of Goodenough's lab. Oxford was on the extreme leading edge of a new field.

But there had to be collaboration. "No one man sits there and spits it out," Goodenough said. "It's through interaction, through our openness to others, where we get an idea."

But such collaboration had to be cautious, as Goodenough would discover.

7

Batteries Are a Treacherous World

After oil prices slid back down from their spike in the energy crises of the 1970s, the urgency went out of battery research. Exxon abandoned electric storage and licensed out Stan Whittingham's lithium battery. Ronald Reagan canceled government-funded energy projects of the prior decade, as did Prime Minister Margaret Thatcher in the United Kingdom.

Japan was different. Though Exxon had distributed Whittingham's lithium batteries in watches in 1977, researchers struggled to build them bigger. Whittingham's work kept igniting, a result of the presence of pure lithium metal as the anode. But, working on the problem for a decade, a Japanese researcher named Akira Yoshino managed to combine Goodenough's lithium-cobalt-oxide cathode with a carbon anode. In 1991, Sony, pivoting off Yoshino's brainchild, released a lithium-ion battery for small electronic devices. Later versions of the Sony battery would contain a better anode made of benign graphite, whose absorptive layers were a perfect temporary burrowing place for lithium ions. But the advance as a whole—the combination of Goodenough's cathode and a carbon or graphite anode—created an overnight blockbuster consumer product. It enabled several multibillion-dollar-a-year industries of small recording devices and other electronics. It triggered copycat batteries and a frenzy in labs around the world to find even better lithium-ion configurations that would pack more energy in a smaller and smaller space.

Despite his central role in the first lithium-ion battery, Good-enough earned no royalties. Unlike Thackeray's South Africa lab, which itself might profit should his invention of spinel prove commercially valuable, Oxford had declined to patent Goodenough's cathode at all—the university seemed to see no advantage in owning IP. In the end, Goodenough signed away the royalty rights to the Atomic Energy Research Establishment, a U.K. government lab just south of Oxford in Harwell, reasoning that at least his invention might reach the market. He never fathomed the scale of the market to come. No one did.

It was not the only time that American battery inventors lost ground in the race to commercialize. Until the middle 1980s, Union Carbide controlled a full third of the global battery market through its Eveready and Energizer brands. But in 1984, thousands of people in India were killed and injured in a gas leak at a Union Carbide chemical plant in Bhopal. In the aftermath, the company sold leading divisions for cash. Its battery unit went to Ralston Purina, which itself ceded lithium-ion to Japan under the rationale that the profit margin per unit was too thin. The nickel-metal-hydride battery that powered Toyota's market-leading Prius was also American born, created by a prolific Detroit inventor named Stan Ovshinsky. After the Prius's 1997 launch as the world's first major hybrid, licensing fees for the battery went to a Chevron subsidiary that acquired Ovshinsky's patents. But Chevron relinquished much of the profit to Panasonic, Toyota, and other Japanese companies that made the final products.

American companies lacked their Japanese competitors' vision, courage, patience, or perhaps all three. Students of economic history ridicule the Japanese juggernaut of the 1980s. They say Japan was a flash in the pan and contemporary panic over its rise a reflection of Western insecurity, not a new, Japanese-led future. But this version of events is not quite right. The Japanese embraced the model of an American celebrated but not emulated at home—Thomas Edison, the consummate tinkerer, who, absent a governing theory to create a new invention, systematically attempted as many ideas as necessary to reach a solution. South Korea and China then also borrowed Edison's method and captured their own large chunks of the global electronics market. As a group, the three countries added energy storage to the arc of America's four-decade-long industrial decline—and a subtext to its anxiety about getting the new battery-and-electric-car race right and dominating the sprawling industries to come.

Charlatans and hucksters abound in eras of invention, since no one can truly know what will become the next bonanza, and batteries have been unusually marked by exaggeration and outright fraud: because people intuitively understand the importance of a much better battery and think that therefore the world should have one, they are vulnerable to deception. In 1883, Edison, misled too many times in the midst of creating his electronic empire, sized up rechargeable batteries as a mere fable. He wrote:

> The storage battery is, in my opinion, a catchpenny, a sensation, a mechanism for swindling the public by stock companies. The storage battery is one of those peculiar things which appeals to the imagination, and no more perfect thing could be desired by stock swindlers than that very selfsame thing. . . . Just as soon as a man gets working on the secondary battery it brings out his latent capacity for lying.[1]

Goodenough tells the story of a Japanese materials scientist by the name of Shigeto Okada. Okada arrived in 1993 at the University of Texas, where Goodenough had moved the previous year from Oxford. He came from Nippon Telegraph and Telephone (NTT), the Japanese phone giant, which requested permission to embed him on Goodenough's team at company expense. After the usual stipulations regarding confidentiality, Goodenough agreed. He put Okada to work next to an Indian postdoc named Akshaya Padhi.

In hosting such researchers, Goodenough was part of the peculiar world of materials scientists, who at their best combine the intuition of physics with the meticulousness of chemistry and pragmatism of engineering. It is their role to dream up a new order from the existing parts in front of them.

Padhi and Okada began to tinker with spinel formulations, searching for one with more energy than Thackeray's manganese spinel and better safety than Goodenough's own lithium-cobalt-oxide. They started by methodically swapping in metals to see if any achieved their teacher's objective. They tried cobalt, manganese, and vanadium, but none was quite right. Finally, they winnowed down the list to a final option—a combination of iron and phosphorus.

Goodenough was skeptical. "Padhi," he said, "you won't get the spinel structure."

Then the old man left for summer vacation.

As had happened with Thackeray at Oxford years before, Goodenough arrived back to news. Padhi said the professor was right—he did not achieve the spinel structure. Instead, he had produced a synthetic version of a different, naturally occurring crystal structure called olivine. And he had managed to intercalate lithium in and out of it. On inspection, Goodenough saw that the result was sensational. Lithium combined with iron phosphate met all the metrics for which he had hoped.

Goodenough didn't learn until much later that Okada—the Japanese researcher—had gone on to disclose Padhi's discovery to his own employer, which had proceeded to secretly develop the formulation itself. In November 1995, NTT, using Padhi's methodology, quietly filed for a patent and began to canvass Japanese electronics makers, gauging their interest in a new, lithium-iron-phosphate battery.

Goodenough caught wind of the subterfuge only the following year. He was incredulous. "Padhi, he was a spy, for goodness sakes," he nearly shouted at his postdoc. "Wake up and start putting something in your notebook." He meant that Padhi should commit his work to writing in his lab book; that record would prove crucial should there be an IP battle. And there very well could be.

"Sorry," Padhi replied to Goodenough. "He is my friend."

A race of priority was joined. The Japanese and the Americans rushed out competing papers and patent applications. On behalf of Goodenough's lab, the University of Texas filed a $500 million lawsuit against Nippon Telegraph and Telephone.

The complications worsened. An MIT professor named Yet-Ming Chiang began to fiddle with Goodenough's idea and filed for his own patents. Asserting that his improvements had created yet another new material, Chiang launched a Massachusetts company called A123. His stated aim was to sell a version of the lithium-iron-phosphate for use in power tools and eventually motor vehicles. This established another legal front for Goodenough as Chiang's company sought to persuade a European tribunal to strike down the old man's patents, which it eventually did in 2008.

The result was a free-for-all, one that reached an apex late in 2008 when Warren Buffett spent $230 million to buy 10 percent of BYD, a Chinese car company that announced a new lithium-iron-phosphate-powered electric car. No one spoke of the source of BYD's batteries but, coming after Chiang's actions, the impression in the industry was that the Goodenough lab's invention might turn up anywhere.

In 2009, A123 sold shares in an initial public offering. Chiang's charisma, the MIT name, and the general tenor of the times created an aura of sizzle, and the share price surged by 50 percent on the first day of trading. Chiang's company raised $587 million, the biggest IPO of the year and a tremendous payday for him and all involved. Except, again, Goodenough.

In the end, the University of Texas settled with NTT. The payoff to the school was $30 million along with a share of any profit from its Japanese patents, recognition that Goodenough had been infringed. Goodenough received nothing from A123. He regarded the outcome as a travesty. The university-hired lawyer was a mere big talker, a naïf out of his depth against cunning shysters. As for the university, Goodenough said it lacked the courage to fight.

8

Creating NMC

I n the early 1990s, the researchers at Argonne's Building 205 were griping openly about oppressive management. The Department of Energy wanted invention on demand but also mandated excessive safety training, the combined impact of which was to "discourage spontaneity." The lab was no longer as secretive—since Argonne was working on so many nonnuclear projects, it had abandoned the practice of declaring everything classified. Much of the work remained confidential, as basic invention was under way, but often did not involve matters of national security. Scientists no longer had to wear color-coded shoes to protect against nuclear contamination. They could take food and coffee into their offices. And their offices were air-conditioned.[1]

Still, you could not enter or move around Argonne without a lab identification badge. They hung by a string from everybody's neck. Many were imprinted with the word "COUNTERINTELLIGENCE." The IDs were mildly jarring in that the photographs often showed a much younger, college-age version of the scientist. In his, Chamberlain resembled a California surfer, with brushed-back hair that could be mistaken for blond. His hair had long since grayed.

Of course it was no crime to brandish a dated photograph, such as Chris Johnson's. He had spent his entire career at Argonne. Now in his forties and fullish, Johnson was once a slim professional with a stylishly trimmed beard. You could imagine the go-getting young scientist who, working with Thackeray as his chief researcher, coinvented Argonne's NMC almost a decade and a half before.

Johnson was an unpretentious and earthy Ohioan. His father taught high school chemistry and strewed science textbooks about the house, but he did not press the subject on the boy. "I just want you to feel like it's not work when you get up and go to your job," he told his son. So Johnson did not at first grow up as a science geek. He did not puzzle over test tubes in the garage or ponder garden insects underneath a microscope. But when he reached high school, a science teacher's enthusiasm infected him, which led Johnson to major in chemistry at the University of North Carolina. There, in the electrochemistry lab, Johnson felt in his element.

In 1991, Johnson joined Argonne as a postdoctoral assistant. Sony had just commercialized lithium-ion.

As he briefed himself by reading scientific journals, Johnson noticed copycat behavior. Papers fixated on the fashion of the day—the Goodenough lithium-cobalt-oxide cathode that had enabled Sony's new batteries. None seemed to pose daring new ideas—they only plumbed how to make lithium-cobalt-oxide better, and even when they did that, their science seemed "lacking." But one chemist stood out—Mike Thackeray, who was working back in South Africa after his Oxford stint. Thackeray was talking about his alternate system—manganese oxide, which he said would cost less than lithium-cobalt-oxide. In Johnson's view, only Thackeray seemed prepared to say something original and produce the data to back it up.

About this time, Thackeray's South Africa bosses informed him that they were shutting down his lithium-ion program. Notwithstanding Sony's coup, the lab did not foresee sufficient sales in the lithium-ion play. Thackeray debated the point, but the decision was made. He was to find other projects.

In 1993, Thackeray met a talkative American named Don Vissers at a battery conference in Toronto. Vissers was a senior manager in Argonne's Battery Department. He and Thackeray agreed that the market for lithium-ion batteries was bound to swell. Yet both were frustratingly on the outside in this discernible trend: while South Africa was erring by abandoning lithium-ion, Argonne was falling behind because of its passiveness in the same field. The Chicago lab continued to work on high-temperature sulfur batteries and had yet to make its own push in the new technology. Vissers suggested that they had a common cause. So why didn't Thackeray consider a move to Chicago and taking Argonne into the science of lithium-ion?

Thackeray pondered it and a year or so later agreed.

Thackeray's wife, Lisa, dreaded moving to an unfamiliar land where none of them—not they or their three daughters—knew a soul. Thackeray described reaching O'Hare that February: "As the American Airlines aircraft approached the landing strip with the wheels a few feet from touchdown, the pilot opened the throttle and took the plane back into the air. There was a stunned silence in the aircraft. Lisa, looking at me at her side, said quietly, 'Thank God—we're going home!'"

They were not going home. The pilot looped back around and landed the plane without incident. Emerging later from customs, the Thackerays saw an Argonne man with a sign. Greeting the family, he bestowed a silver dollar on each of the daughters. The gesture swept aside Lisa's apprehensions about life in a new country.

Work started at once. Thackeray adopted Chris Johnson as a protégé and took him along to an international lithium battery conference in Boston. Arriving there, Johnson watched as a slew of scientists greeted Thackeray in the hotel lobby. "Everyone knew Mike," Johnson said. "Everyone was coming up to him. 'How are you doing? I understand you are at Argonne now.' I am thinking, 'Wow, he is really major in the field. And this is going to be a really nice relationship.'"

Thackeray began to brief Johnson about his plan. If you reduced the amount of expensive cobalt in the cathode and substituted plentiful manganese in its place, you could make batteries that were both cheaper and safer than Goodenough's industry-standard chemistry. But you could only use so much manganese because it tended to degrade over time and destroy the battery's performance. Instead, you needed to deploy it together with nickel, which preserved the manganese and hindered its degradation. That made the ideal compound a combination of nickel, manganese, and cobalt, or NMC, coupled of course with lithium.

Yet while this formulation was striking, it did not break new ground. The problem was that physics stepped in and spoiled Thackeray's picture. Nickel, manganese, and cobalt, it turned out, would come apart just like Goodenough's formulation if you sent too much lithium into the shuttling motion between electrodes that created electricity.

Thackeray thought back to South Africa. He had learned that a compound of lithium, manganese, and oxygen that went by the atomic lettering Li_2MnO_3 was electrochemically inactive. It was normally cast aside as an impurity. But now Thackeray's intuition told him the story was incomplete—he thought there could be more to the material than anyone knew. His idea was to add a bit of Li_2MnO_3 to the lithium-laced NMC. Thackeray suspected that this twist would buttress the NMC and keep the cathode intact as the battery was charged and discharged.

In 1994 and 1995, Johnson created test battery cells using the formulation that Thackeray described and intercalated the lithium. He found that he was able to shuttle well over half the lithium between the two electrodes, all while the NMC structure held very much together. It was as though the cathode had been waiting for the Li_2MnO_3 to provide it stability.

Johnson learned why Thackeray's intuition was correct. Even though the Li_2MnO_3 was itself inactive when introduced into a cathode, its manganese and lithium went on to migrate and lodge in the NMC like pillars. These atoms propped up the structure while the lithium in the NMC began to shuttle.

Visually, both NMC and Li_2MnO_3 resemble a stripped-down

house. The floor and ceiling are made of oxygen atoms, and the walls comprise nickel, cobalt, and manganese. Scientists call this framework a lattice. Because the lattices of the NMC and the Li_2MnO_3 are similar, Johnson could easily integrate the two at the nanoscale.

If the only notable thing was that the compound now held together, Johnson would have been engaged in a mere thought exercise. But stability wasn't their only success. If you were thinking about an electric car, the NMC led to a better cathode than Goodenough's lithium-cobalt-oxide, his lithium-iron-phosphate, or Thackeray's own manganese spinel. Not only was it cheaper and safer, but Thackeray also calculated that the extra lithium in the system improved its performance. The double lattice let you pull out 60 or 70 percent of the lithium before collapsing, well over the 50 percent you could withdraw from Goodenough's lithium-cobalt-oxide. That extra lithium—the added 10 or 20 percent—meant more energy.

Thackeray called the invention "layered-layered," or "composite."

This double lattice had another advantage. It set up Thackeray for future advances. He could swap other metals in and out of the latticework to make more improvements.

As it was, though, the NMC was already potent. It overcame an essential challenge facing batteries if they were ever to compete against gasoline propulsion, and that was that very few people would settle for a single trait in an electric car. The ability to travel a long distance was important, but it was not sufficient; drivers demanded other qualities, too. They wanted the car to take off— immediately—when they pressed the accelerator, and to keep on accelerating to high speeds. They insisted that their vehicle be safe—consumers, not to mention regulators, would reject any car with a chronically explosive battery. The last quality was possibly the hardest to deliver: pushing for such performance in distance and acceleration tended to make the battery more dangerous.

Cars equipped with Argonne's NMC formulation could travel forty miles on a single charge, a key technological marker because it was the distance that the average American motorist drove in a day. If you did not meet that metric, you couldn't really think about putting a model on the road. The NMC also provided the rapid acceleration demanded by Americans. And manganese made the system safe.

All in all, NMC was superior to any cathode thus far produced in the national laboratories, even, some said, to anything designed elsewhere.

The breakthrough bucked up Thackeray, who seemed endlessly curious tinkering with the narrow range of elements on the periodic table relevant to batteries—but only if he intuited the potential for a meaningful advance. No one could predict commercial interest, which often seemed unfathomable. Why, after all these years, did Goodenough's lithium-cobalt-oxide remain the standard lithium-ion formulation, used in virtually every cell phone, tablet, and laptop on the planet? No one else's work—even Thackeray's spinel—had been good enough to eclipse the old man. That illustrated the extremely slender chance of commercializing something new. Still, there had to be the chance of outdoing Goodenough— of progressing toward the ultimate goal, which was challenging the provenance of the internal combustion engine. Otherwise Thackeray was not interested.

He began to assemble a patent application for his NMC.

In May 2000, Thackeray flew to Italy's Lake Como for a two-week lithium-ion conference. The setting was symbolic—Alessandro Volta was born in Como in 1745. Just eight months earlier, the city had hosted the bicentennial celebration of Volta's invention of the battery. Some two hundred experts from thirty countries had gathered to mark the occasion. But Thackeray was disappointed to find little feeling of the past at the May event. For starters, it was held at a conference center a train ride away from the city, where he found the atmosphere sterile.

Thackeray delivered one of the opening presentations. Midmorning the next day, he sat in on a thirty-minute talk by four scientists from New Zealand. In excited language, the men spoke mysteriously of a new approach to batteries coupling chromium with manganese oxide. Later, Thackeray strolled by a poster display manned by one of the New Zealanders, a crystallographer named Brett Ammundsen, and found him frustrated. "You of all people will know what I'm doing," Ammundsen said. Surely Thackeray, the pioneer of manganese spinel back at Oxford, grasped the significance of the New Zealand advance even if no one else at Como seemed to.

At that moment Thackeray *did* comprehend what the New Zealanders were up to: treading on his turf.

For Thackeray, they were uncomfortably close to his maneuver with Li_2MnO_3—just as he was, they were injecting added lithium to juice the performance of a cathode, in their case a chromium-and-manganese oxide formulation.

An alarmed Thackeray telephoned Chris Johnson in Chicago.

"Quick, do a couple of more experiments and then write up an invention report," Thackeray said. "We are going to file for a provisional patent." The patent he had been preparing was not quite ready. But now it needed to be if he was to get the jump on the New Zealanders.

A "provisional patent" was a tactical move—it was what you filed when you had confidence in your idea, were in a race with rivals, but lacked sufficient data. It provided a full year to validate your claim. If you found your data, you could be awarded a full patent, dated when you originally filed. Johnson dropped what he was doing and went to work. When Thackeray arrived back in Chicago, they both began to produce test cells and create electrochemical data that more or less validated their claim to greater performance. They sent off the data to the lab's outside lawyer. In a diagram, Thackeray claimed broad priority for a cathode combining nickel, manganese, and any third metal. A year later, they filed and were awarded the permanent patent.

They had beaten the New Zealanders. But Thackeray needn't have worried. It was a long six months after Lake Como before the

New Zealand group filed its own patent application. Reading it, Thackeray found it mediocre. The New Zealanders had "missed the big picture," he thought. It was as though they did not understand that the key to the material was the interaction of the two lattices—the use of the Li_2MnO_3 to stabilize the NMC structure. Rather than a composite of two structures—the central fact of the formulation, Thackeray posited—the New Zealanders thought it was a homogenous mishmash of metals.

So it went with another competing patent application that surfaced at Dalhousie University in Halifax, Canada, one that to Thackeray also seemed confused. "They just didn't know what they had," he said.

To Thackeray's eye, the other applications resembled "what the Japanese will do," which was "claim the world." If you did so, without describing precisely what you meant, you could be ripped apart by patent challenges. The trick was to be clear and simple so that there was no mistake about what you were asserting. "They sort of meandered along a road, blind around every corner as to where they were going," he said, an approach that left you "running into problems with priority." Thackeray's own exacting method had been passed down by British mining executives in South Africa, the fellows who expressed commercial interest in the Zebra battery and who would take him and other battery scientists for lunch in a bar "to brainstorm" how to protect it. Patents would be drafted with beer and wine flowing freely. Thackeray noticed that the South Africa patents tended to hold.

A few years later, when carmakers began to produce electrics, they were highly secretive about their batteries, including which formulation they used and the cost, regarding such knowledge as competitively vital. But General Motors openly announced that it had bought licenses for both of Thackeray's major inventions—the NMC and manganese spinel—in a combined-formulation battery for the Volt, its first new electrified car, a plug-in hybrid that it launched in 2010. GM said the battery's forty-mile distance was ideal for a first-iteration Volt. But GM's interest was not confined there; Argonne had promised an advanced version of the NMC, one that could be combined with an improved anode and take the car much farther. GM was waiting for that advance, with which it hoped to launch new electrics.

The addition of Thackeray had injected much-needed competitive verve into Argonne's Battery Lab. At once, it was on the leading edge of lithium-ion research. The lab's next recruit would complete its special tandem—a pair of battery men who sat astride both the scientific and commercial worlds.

9

The Man from Casablanca

The Moroccan village of Benahmed is a quick half-hour drive down a smooth highway from Casablanca. But when Khalil Amine was growing up there in the 1960s and 1970s, the trip took twice as long, winding down narrow roads on a bus. Benahmed was a clean, bright town with a small French population that stayed on after the end of colonial rule a few years before. Amine's father, an Arab intellectual who taught school, and his mother, a Berber, produced seven boys. Khalil was the second. Of his mother's family, Amine said, "The Berbers are extremely good in business."

Family lore went back to the first decade or so of the twentieth century, when Amine's maternal grandfather, Benadir, was a twelve-year-old shepherd in the mountains around the port of Agadir. Pretty often, an elderly man would beat him. But one day, a provoked Benadir took a rock and smacked the old man across the head. The man fell and did not move. The terrified boy fled.

Benadir was hiding when a fruit cart attached to a tractor trundled by. He clambered aboard and quickly concealed himself. For the next two or three days, the cart progressed up the coast. Benadir feasted on the fruit and vegetables. But in Casablanca the fruit sellers discovered their stowaway and threw him into the street. Benadir began to walk and beg. Tired and dirty, he turned up at a home. A Frenchwoman inside, filled with pity, took Benadir in. She cleaned him up and allowed him to stay on as a house-keeper.

Amine does not recall the woman's name, but one day, went the story, her businessman husband asked Benadir to mind one of his shops. This was a tremendous responsibility, as Benadir saw it, and he dutifully opened at five A.M. and closed at midnight. He slept and ate at the shop. After a while, the Frenchman observed that the shop's earnings had soared. He assigned the boy more shops and began to treat him like a son.

In 1956, Morocco won independence from France and Spain. Benadir's French family was among a mass of panicky foreigners who repatriated. On his way out, the Frenchman offered his enterprise to Benadir. So it was that Amine's grandfather became a considerable local titan.

The stories may have rubbed off in Amine's own commercial instincts—"a gene from my mother's side, I think," he said. "But unfortunately, the twist is that Benadir has a son who never went to school. He never worked. He destroyed that fortune. My uncle, yeah. If I were him, I would be owning half of Morocco." Amine was laughing but he wasn't joking.

Amine said, that, as a boy, he was serious and emotional—he would cry if a classmate outscored him on a school test. At a French boarding school outside Casablanca, where he was sent since his own village had no high school, Amine and some friends read late into the night and early in the morning in the bathroom, the one place that the masters left lit. Mostly, his high school friends played soccer, smoked, and played cards. Amine sat with them, reading. In the case of a disputed game, Amine would arbitrate. "I could see who was the winner and say, 'Okay, here is the money.'" He was their trusted person.

As a college senior, Amine topped Morocco's national science exam. That qualified him for a fellowship at France's University of Bordeaux. There, he earned his Ph.D. in chemistry and accepted an offer to become a postdoctoral assistant at Kyoto University in Japan.

· · ·

In Kyoto, Amine moved into a university dorm that to him seemed like a stylish hotel. When he wasn't in the lab, Amine watched TV in the dorm lounge, and that was where he was when an extremely attractive young woman sauntered in and took a seat. "She was like a movie star," Amine said. "Wow. Blow me out." He started to chat. "Where are you from?"

Her name was Xiaoping Xu. She was Chinese and had been in Japan for three months. Just now, she was preparing for a medical school exam that included technical Japanese. "I don't have books," she said, "so I'm worried."

Amine said he had books and dropped them off with her. Then Xiaoping vanished.

About six months later, he received a note. Xiaoping had finished first in her class and won a prestigious pharmaceutical fellowship. She wanted to return Amine's books.

Amine was not thinking books but serious romance—he genuinely liked Xiaoping. He thought she must be interested, too, but a Chinese friend told him that, if she was as conservative as Amine described, he had to slow down. He could not be pushy.

He invited Xiaoping on a series of expensive dates—to dinners, to temples. He bought her gifts. On Amine's mind all the while was a two-decade-old memory from Morocco. He was six and along with his brother was watching an action movie starring two beautiful actresses: a Chinese and an Indian. "I'm gonna marry either that Indian girl or that Chinese," Amine had said. Now a grown man, Amine wanted to realize that early childhood fantasy. He was intent on marrying Xiaoping.

After six months of courtship, Xiaoping allowed him a kiss.

In the lab, too, Amine felt a long-won sense of success. He was a researcher who did not necessarily find original pathways to store more atoms in smaller spaces, but he read voraciously, paid rapt attention at conferences, and could rapidly grasp both the potential and the flaws in ideas advanced by others. He would e-mail friends with questions and thoughts and from there identify creative solutions before anyone else. "If there is a problem, we fix it," Amine would say. "If there is another problem, we fix that one, too." He had moved on to become a research and development manager for the Japan Storage Battery Company, a privileged position for a young foreigner in the East Asian country. Japan was booming and the money flowed in high salaries and astronomical extras: double overtime, which would mean triple salary; an extra six months of salary each year—three months for the winter, three for the summer—and additional bonuses for key players who made important breakthroughs. Employees were not paid royalties because the company retained rights to all inventions. But Amine was awarded an added bonus equivalent to double his annual salary for his invention of a five-volt battery system using nickel and magnesium. He received another two-year bonus for a cobalt-oxide system that the company licensed to Sony and Samsung, a link back to John Goodenough's blockbuster original.

Amine and Xiaoping decided to wed. Amine's parents immediately embraced the idea. As for Xiaoping's family, Amine had to impress her mother. In Chinese tradition, "if the mom says it's okay, everything's smooth," Amine said. "If the mom says no, you are in trouble." At first, the signs were not good. Xiaoping's mother thought that Amine, given his professional success, must be advanced in years. "Why do you want to be with this guy? He's old," she told her daughter. Amine said, "She thought I was like fifty years old." But just three years separated him from Xiaoping.

They traveled to China. Before going, Amine checked in with his Chinese friend. It went without saying that Amine would present a gift to the mother, but it had to be valuable, the friend said. "You bring her a flower, and they'll joke about you. They'll say, 'Yeah, this is a dud.'"

Amine arrived with a gold pendant-and-bracelet set, several French scarves, and a pair of fashionable shoes.

"Hello," he greeted Xiaoping's mother, smiling with gifts in hand. She smiled back in a way that told him there would be no problem.

In 1997, Xiaoping was accepted to medical school at the University of Illinois, a necessary step if she wanted to practice in the United States. She encouraged Amine to follow her and he found a job leading a battery research group in Ann Arbor, Michigan. It wasn't far—he would drive the four hours to Chicago and see Xiaoping as often as he could.

One day, Don Vissers walked into Mike Thackeray's office. He was just back from a conference in Oslo. Did Thackeray know of Khalil Amine, a battery guy in Japan? They had shared lunch and Vissers seemed enthused. Thackeray said that, yes, he did know of Amine and was also impressed.

Vissers soon found himself in Kyoto. Amine showed him around his lab.

"God, don't tell him too much," Amine's boss had instructed. But over dinner that evening, Amine went on about the science.

"Would you be interested to come to the U.S. to work with us?" Vissers at last asked.

"Sorry," Amine said. He was moving to a job in Ann Arbor.

Vissers laughed.

"Your wife is studying only six miles from Argonne," he said.

"Really?" Amine replied. He had equated Argonne with the 1940s atomic bomb tests and assumed it was nearer to New Mexico.

"Count me in," Amine said.

Amine and Xiaoping, now married, used their savings from Japan for a house in Downers Grove, the middle-class "Little Argonne" to which the lab's first scientists migrated in the late 1940s. When she finished medical school, Xiaoping opened a holistic medical clinic nearby. Her instinct of shifting to the United States had been perfect, Amine said—"for the kids, and also for my career and her career." Then Xiaoping found a foreclosure a few miles away in a newer, wealthier subdivision called Oak Brook. The schools there were terrific and progressive, too. They took the house.

Amine might have continued to be successful in Japan, but not Xiaoping. There was too much prejudice against Chinese, not to mention bullying of foreign children. In Oak Brook, Xiaoping was happy and popular.

Xiaoping's parents moved in with her and Amine. When Amine arrived home from a long business trip, regardless of the time, he would notice someone in the window. It would be Xiaoping's mother. "She is very worried. You know, about me," Amine said. But when he opened the door, she would be gone—in her room, falling silently asleep.

10

Theft in the Lab

Don Vissers walked into Thackeray's office.

"Khal thinks you are stealing his ideas," he said.

What?

Amine, livid, had griped that Thackeray's application to lock in the provisional NMC patent had pilfered his work. Two years after his recruitment by Vissers, the Moroccan had become a force in the Battery Department. He arrived at the office at six A.M., before almost anyone else, for a head start on funding applications to the government and private companies. These project proposals demonstrated a knack for ferreting out the potential next big thing, connecting the hidden dots in raw work under way in the field and adding the necessary missing ingredient. Many of the applications were approved and the money was on course to make up two thirds of the entire departmental budget, with the corresponding proportion of staff under his direct supervision. At root was Amine's drive, a disruptive ambition that made much of the department seem to be standing still and upset many of his colleagues, who called him an opportunist who exploited others' work.

Amine and Thackeray themselves were an incongruous pair, not only because two immigrants from opposite ends of Africa were driving Argonne's effort to push American dominance into the next industrial era. Thackeray—gently spoken yet typically disheveled in an ill-ironed shirt, his hair combed straight down—was fastidious about his science and fanatical about understanding events at the atomic level. He did not quite comprehend Amine, who, while impeccably dressed and coiffed, was by comparison a brawler, principally interested in what he could push into the marketplace. Thackeray was baffled as to why Amine did not seem bothered by the details of a given compound's behavior. But for whatever reason Amine didn't, and Thackeray shook his head and focused back on the science.

The allegation that Amine leveled against Thackeray was toxic. You could accuse a scientist—a *colleague* on your own team—of almost nothing worse. If there truly was a theft, Thackeray's image would be blackened. But if Amine was wrong, his credibility would suffer.

That was not all. Amine had also accused a supervisor of bigotry. The manager, Amine said, treated him and his Chinese lab assistants as "third class" members of the lab even though they were among its most productive scientists. If the supervisor didn't change—fast—he was going above his head. This Amine did—he went to a more senior manager and walked out with a promotion.

Amine's accusation was unfair—the manager was no bigot. He was simply put off by Amine's aggressive style. But Amine's charge against the extraordinarily civil Thackeray crossed a different line. It staggered and mystified the South African. Amine went so far as to forbid his staff to speak with Thackeray.

Chris Johnson thought he knew what was going on. One day, he had been conversing with a South Korean named Jae-kook Kim, one of Amine's postdoctoral assistants. "I have this theory about titanium—that you can add a little bit of extra lithium to it" and produce a higher-capacity battery, Kim said. It was the same thesis as Thackeray and Johnson's but advanced the use of titanium rather than manganese. Johnson reckoned that Amine had learned of the breakthrough and, without overtly saying so, was proposing a twist. Typically, he had found a way to grab a piece of a big, new idea. Johnson wasn't ruffled—he thought titanium and manganese could be a good marriage. You could claim both approaches in your patent application and make it even stronger.

But that was Johnson, with his decorum-driven Midwest sensibilities. Thackeray, with a cultural spine anchored elsewhere, felt differently. He was seriously aggravated.

Amine seemed bent on warfare. That left it to Thackeray to conjure up a resolution—if he chose to. The thing was, legally speaking, you must justify your name on a patent. It was not like a research paper, in which anybody is eligible for a byline. When it came to a patent, you had to genuinely contribute something, specify what it was, and assign to it a percentage of the whole work.

Thackeray considered these questions. There was no collaboration between his and Amine's research groups—no one was working together. So there could have been no direct blurring of ideas. Amine's group *was* adding lithium to the formulation. He had swapped titanium oxide for Thackeray's manganese oxide in order to achieve higher capacity. But his premise was different—like the New Zealanders, Amine did not describe the double lattices. Thackeray thought Amine had missed the big picture and that his ideas were at best peripheral to the actual invention—"a tiny, little thing," he said. Amine did not belong on the patent—period.

Yet Amine was not backing down and Thackeray felt the heat. "It was his word against mine," he said. Thackeray could simply put his foot down but then the fight was sure to become uglier.

Thackeray opted for peace in the lab. He added Kim and Amine to the list of inventors. The advance reflected by NMC was bigger than their personal differences. But a pall went over the relationship. You couldn't make that sort of allegation and expect others to forget.

The NMC patent attracted attention. At a Boston battery conference, "people were just taking pictures of my presentation and taking notes," Chris Johnson said. But Thackeray recalled one of John Goodenough's maxims: once you have an invention, you have a two-year lead time before other scientists catch up. In the case of the Argonne group, it had an edge in a new, powerful, and cheap battery system. But with the patent filed, they had to be alert.

In 2010, Thackeray and Johnson were startled by a report from Dalhousie University. Jeff Dahn, a blunt and outspoken battery researcher whose own version of the NMC had been patented by the 3M Company just after the Argonne pair, announced a big jump in the material's performance. It happened when, as an experiment, he juiced the voltage. The capacity surged.

If you pack lithium into a battery and apply voltage to move it from the cathode to the anode—the act of charging the battery— the structure puts up fierce resistance. It restricts the lithium's free movement, thus limiting how fast energy can be extracted, and thus how fast a car could go. Some goes astray along the way, stuck in one or the other side of the battery. In the case of NMC, it had high energy—you could pack in a lot of lithium—but relatively low power, meaning that you could not extract the lithium very fast. What Dahn did was to raise the voltage used to charge the battery above 4.5 volts—to about 4.8 volts, considerably more than the usual 4.3. That boost triggered a race of shuttling electrons. The result was staggering. Theoretically speaking, Dahn was putting almost all of the lithium into motion between the cathode and the anode. In principle, you should not have been able to extract that much lithium from the cathode, thus removing important walls from the latticework of the cathode—the house of oxygen and metal atoms should collapse. But Dahn discovered that he *could* do so.

Johnson went into the lab and tried to duplicate Dahn's claims using the Li_2MnO_3. He pushed the voltage over 4.5 volts. Just as Dahn had reported, the capacity surged.

It was an important discovery. The numbers told the tale. Ordinarily, lithium-ion batteries such as Goodenough's lithium-cobalt-oxide store around 140 milliampere-hours of electric charge per gram, a revolutionary capacity when it was invented but insufficient for the ambitions of the new electric age. By pushing the voltage, Johnson was getting much more—250 milliampere-hours per gram, which was even higher than the 220 that Dahn was reporting. Trying again, Johnson got 280, almost twice lithium-cobalt-oxide's performance. The experiments suggested that the NMC was even more powerful than they had thought on pioneering it five years earlier—far more. At once Li_2MnO_3 was not simply a fortifying agent, as had been presumed. At just over 4.5 volts, it came alive in a very muscular manner. At this higher voltage, you activated a new, heretofore unrecognized dimension of NMC. This was NMC 2.0, the breakthrough that could push electric cars over the bar and challenge gasoline-fueled engines.

The Argonne men published their own results immediately. As for the IP, they were covered—the jump in capacity at higher voltage was simply a new understanding of the original 2000 application.

Working in the lab with his own team, Amine made an additional advance. It was in a usually overlooked part of the battery—the electrolyte in which the cathode and anode are submerged. It is this liquid that allows ions from the anode to migrate to the cathode, and vice versa. But sometimes the battery becomes overcharged, creating the risk of fire, a phenomenon that had already inflicted considerable public relations damage on lithium-ion batteries and products containing them—you could have only so many laptops burst into flames in airport lounges and elsewhere before consumers began to worry. Amine's team invented and patented a new molecule based on boron and fluorine that, when added in powder form in minuscule amounts to the electrolyte, absorbed excess electrons and thus reduced the chance of fire. Amine was a full-fledged member of the NMC team, a handful of researchers who had now expanded their work into a constellation of patents centered on the NMC that was arguably more valuable than any rival new battery work.

11

The New Boss

Not much more came of the NMC until the arrival of Jeff Chamberlain at Argonne in 2006. Chamberlain was tall, muscular, and relaxed, with small hands for his build, used in controlled, almost robotic gestures that conveyed confidence without seeming to be showy. It was his voice that captured attention in meetings. In a room of competing opinions, his basso profundo seemed to prevail. The voice made it impossible to ignore Chamberlain when he began to moralize. Among his gripes was "anti-intellectualism among elected officials." Another was how Americans were "beholden to the interests of those who produce oil." Chamberlain would continue to anyone listening: "We are the Saudi Arabia of coal and have nuclear energy. We should aim at energy independence with coal, solar, wind, and nuclear, then use them to charge up electric cars. Use electricity instead of oil—for everything. How do we get there?"

He was hokey, which endeared him to the rank and file, scientists who were unmoved by talk of a battery war but gung-ho on the subject of importing less Middle East oil. Their passions rose at the idea that batteries could help stop climate change. They believed Chamberlain when he said over the following years that many oil despots would be in trouble if drivers turned to electric cars to the degree Obama and Wan Gang both sought and those vehicles were charged with electricity produced by natural gas. Oil prices would fall, undercutting the long-running flood of money to

Russia and OPEC, especially members that themselves did not possess gas. Since China would require less foreign oil, a current subtext to tension with outsiders—its colossal need for imported resources—would soften, and its air would be cleaner. When you added up these factors, you also emitted much less carbon. What was to dislike? Chamberlain understood that his boosterism infused the lab with a sense of purpose and that led him to promote the big energy picture even more.

Chamberlain grew up in Longwood, a small town near Orlando where his father, Jack, sold marine engines and family conversation often turned to the auto industry as his grandfather and uncle both worked for Ford Motor. Recalling those years, Chamberlain would tell of knocking on neighbors' doors with his Red Rider wagon, collecting newspapers to recycle. He and his father would weigh and tie the papers in twenty-five-pound bundles that reaped twenty or twenty-five cents each at the recycling center. Jack Chamberlain explained the commercial chain to his son—idea, execution, money, personal benefit. The son was given trombone lessons as well and by high school was good enough to join a brass quintet. In addition to him, there were two trumpets, a French horn, and a sax. Before long, the group qualified for state competitions and played a gig at Disney Village over in Orlando. In his senior year, school band and chorus members rallied together and helped to elect Chamberlain as class president. It was quite a coup as his opponent was the school quarterback and an unforgettable cap to his years in Longwood.

He had meanwhile won a full scholarship to Wake Forest University and following that earned a Ph.D. at Georgia Tech. Chamberlain had expected to teach. But in 1993, when he began to search for university-level positions, he found that much had changed while he was at school.

Back in 1982, a federal judge had ordered the breakup of AT&T, the telephony monopoly. Bell Laboratories had been AT&T's research arm, and when it divided into parts in line with the court order, Bell eventually fell apart, too. By the early 1990s, the lab had shrunk. Other industrial research centers followed the same path—General Electric, RCA, and Xerox also diminished their basic research units, firing and retiring thousands of experienced researchers who now poured onto the job market. So many first-rate scientists were available for hire that they all but shut out the prospects for young, freshly minted Ph.D.s like Chamberlain.

There was work, however, if you were open to less-fashionable industries. Chamberlain accepted jobs at a series of mining and chemical companies where he carried out tasks such as leaching gold and copper from rocks and developing new semiconductors. It was not long before his employers perceived a latent talent—an unusual ability to speak to anyone as an equal. To them this meant potential in sales. But if Chamberlain was possibly more valuable on the sales staff than in the lab, his raw ability would require honing. In his first job, his bosses began with instruction in deciphering a client's desires, in addition to what he didn't *know* he needed and wanted. Later, Chamberlain would recall those lessons quite a bit.

Chamberlain's closest friend in these years was Dave Schroeder, a smart, funny, and mouthy Illinois native with whom he worked at a microprocessor company called Cabot Industries. Neither man was liked much by the Cabot hierarchy, which regarded them as troublemakers. For their part, Chamberlain and Schroeder sensed capitalism gone awry with managers who created an unnecessary "battling culture." They felt they might do a lot better if they started their own firm. So, while keeping their day jobs, they tried out some ideas.

Their first brainchild was Chamberlain's. It revolved around fantasy baseball. For those unfamiliar with fantasy sports, it is a multibillion-dollar-a-year industry[1] whose participants choose imaginary rosters of players from real sporting teams, and keep score—often for money—based on the players' individual statistics. Chamberlain devised software that he thought could improve the chances of winning. It borrowed from rolling averages used by stock pickers, the insight that events tend over time toward the average. Consider baseball hitting—"somebody like, say, Ryan Braun," Chamberlain said. "He plays for the Brewers and was the MVP of the National League. He's twenty-eight, twenty-nine now, and has swung a bat since he was five." Say Braun's hitting suddenly takes a dive. "The fantasy baseball player would say: 'Oh God, the batting average is way down; I am benching him.'" But Chamberlain's thesis was that that was precisely the moment to muster the nerve to keep him in. Because Braun would return to his average. And the only way to do so was to hit above his average. "It was counter-intuitive," Chamberlain said.

The idea was to sell it to Yahoo!, which at the time had hundreds of thousands and possibly millions of fantasy sports subscribers paying ten dollars a month to monitor statistics in real time. Calling the invention "Trend Tracker," Chamberlain and Schroeder filed for a patent.

One day, they sat down for a meeting with Yahoo!'s senior fantasy sports executive.

"I can't believe we didn't think of this," the executive said. A young assistant said, "Okay, if it is that great, let's *us* do it. Thanks for the meeting." The executive glared. "Did you hear at the beginning of the meeting when they said they already filed patents on it?" he said.

He said he was prepared to do a deal with Chamberlain and Schroeder. "We'll figure out a way to make this happen," he said.

Two weeks later, the man was fired for unrelated reasons. His successor didn't warm to Trend Tracker.

Chamberlain and Schroeder tried another idea. A material known as a dendritic polymer was generating excitement. It was a compound that could be turned into a variety of products. What caught Chamberlain's and Schroeder's attention was that it could prevent melting in silicon wafers, a crucial need in computers—you needed to remove as much metal as possible and keep down the heat or your system would go down. A New England inventor had found a way to make dendritic polymers cheaply, and Chamberlain and Schroeder took his idea to Silicon Valley. Here was a certain path to fortune. But no venture capitalist they met felt the same confidence. All the pair heard was, "Do you have anything in energy?" The issue was timing. The smart money was shifting from chips to alternative energy.

The start-up failures coincided with a fiasco in their day jobs. The Cabot bosses created new employee performance evaluations based on general personal goals—something about "you know, honesty and integrity," as Schroeder put it. A group meeting was held to select five such goals. Both Chamberlain and Schroeder voted against "excellence" as an objective, "the dumbest thing I'd ever heard," an immeasurable metric, they thought. To illustrate its inanity, Schroeder nominated what he considered an equally unquantifiable goal—"courage." Chamberlain and a bloc of colleagues voted with him. They won. An unamused boss marched up. "Do you want to work here?" he shouted.

Chamberlain resigned. Awhile later, Schroeder did too.

With three children, Chamberlain needed a new paying job. Start-ups could be attempted on the side. Only one place in the Chicago area was still looking at nanotechnology, the science with which he felt most familiar—Argonne's Center for Nanoscale Materials. But when he applied, Argonne managers assessed his résumé differently—like his previous bosses, they saw a salesman.

The new Washington credo called for an alignment of government and business. Federal lab managers had a new mandate to "do science in a way that was relevant" to industry, Chamberlain was told. He clearly possessed risk-taking entrepreneurial credentials. Thus, he could help arm Argonne for survival in the new political environment, precisely where it most needed help.

The lab offered Chamberlain a position in its intellectual property unit, the office that sought to license inventions to companies. He would handle the licensing of battery patents, which seemed ripe for exploitation but bafflingly were going unnoticed. Chamberlain accepted. But he told Schroeder, "I'm not actually going to do anything. The performance of this group is terrible. No one will notice if I don't perform."

Generally speaking, the traditional role of the national labs was fundamental research, the type that moved science forward but might or might not be commercially applicable. But things were changing. Argonne and industry started with an existing symbiotic relationship. With American research labs dismantled, companies were looking for others to conduct this basic work while they concentrated on design, manufacturing, and marketing. Argonne took the long view and saw itself as a throwback to the old private labs. But it had no commercial profile—it didn't scale up or manufacture anything. It required relationships with companies to put its advances in front of people. This mutual need was now reinforced by politics: one would invent and the other manufacture and sell.

Chamberlain understood this symbiosis intuitively. Schroeder noticed that he began to mention meetings with companies. "Hey, do you remember when you took this job and you said you weren't going to do anything?" Schroeder said.

"Yeah, I tried," Chamberlain said. "But I guess I'm working now.'"

For a year, Chamberlain familiarized himself with the Battery Department's patents. He read a thirty-page primer on the patent portfolio drafted for the lab by Ralph Brodd, who at eighty years old was probably the country's preeminent private consultant on lithium-ion batteries. Brodd's paper singled out the lab's NMC formulation as a particularly attractive property. Still, the invention only truly captured Chamberlain's attention when, around industry conferences, he kept hearing the representatives of major Asian companies boast about their ingenious adaptations of NMC to suit consumer desires like durability and safety. Chamberlain felt he was "opening a box and finding gold." But what he did not hear was any of these companies taking note of Argonne's *patent* for the chemistry.

The lab's problem was that it lacked international patents for the invention. Seeking foreign rights could raise patent application fees to well over $100,000, especially if you filed in China, Japan, or South Korea, the places where, because of serious competition or thievery, one required protection. Much of that was legal fees, since the patent application was certain to be challenged on the first go-around. The patent team would not agree to international protection for every invention—it would be too expensive. So to save money on the NMC, the intellectual property unit had filed only for the American patent.

The simplest explanation for the lapse was that the patent team was not fully attuned to the coming international frenzy over batteries—the lawyers simply did not and perhaps could not know how big the market might become and just how unique the NMC could be within it. For the same reasons, Thackeray himself had not urged international filing.

Thackeray regretted it. You could detect the unhappiness in the battery group. In their view, it was one of those spectacularly shortsighted blunders that you could avert only by patenting everything, everywhere. Now the stuff was in the hands of numerous Asian companies that were either licensing a competing cathode manufactured by 3M, which held international patents, or, in most cases, simply availing of the NMC for free. Argonne—like John Goodenough with both of his big inventions—had not earned a dime.

Chamberlain was determined to corral back some of the lost payoff. It would not be simple, since no company would easily pay to license a product that it currently used for free, and those were the companies he was going after—those brazenly using the NMC with no license from anyone. One approach, of course, was a simple appeal to honor—Chamberlain could try to persuade one or more of the Asian players to buy a license out of fairness to Argonne. If that succeeded, other companies would probably take notice and, in order to avoid a potential lawsuit, pay for a license, too. But he doubted the strategy would work—what is fair in Asia, which seemed to him the most cutthroat business environment on the planet?

Here is where Chamberlain's industry experience came in handy: he decided to create an artificial shortage of the NMC. The idea was to offer the sale of two—and just two—worldwide licenses to make the NMC. If anyone outside the United States wished to use Argonne's NMC in their products, they would have to buy it from one of the lucky pair of companies, which, on purchasing the rights from the lab at a price to be set in a negotiation, would divide worldwide, non-American privileges to manufacture the material between themselves. Once those were dispensed, there would be no more such licenses. The result, he hoped, would be a suddenly excited market for the technology. It was a devilishly misleading strategy, since Argonne did not own international rights. Chamberlain thought it might just work. Anyone could gamble and manufacture an NMC-equipped battery technology without a license. But if they did, they could be assuming a big risk—one or both of these license holders were likely to sue.

It was a gutsy call for Chamberlain. What if the market simply ignored him?

Chamberlain had the advantage of Thackeray's and Amine's prestige. Over the years, the senior managers of virtually all the main manufacturers around the world had passed through and met with either or both of these battery stars. The pair introduced Chamberlain to the industry and he launched a global tour of companies. Two companies bit—the German chemical giant BASF and the Japanese chemical maker Toda. In separate 2008 ceremonies, Chamberlain signed them to agreements, among the largest in terms of cash in the lab's history. Now there was a world market price for Argonne's NMC, and the lab was earning money on a portfolio for which previously there had been none. If you wanted the NMC, and were normally averse to risk, you had to pay for it. That new fact had a measurable impact on the NMC's desirability—Chamberlain's phone began to ring. It was the same companies that had declined his NMC pitch. They had assumed he was bluffing. "How much did you say a license costs?" they would say. But he wasn't bluffing—the two sole global manufacturing licenses were taken. There would be no more. Now they were upset. Chamberlain's artificial shortage had turned into a genuine one.

As for the United States—no one was transgressing the Argonne patent for the simple reason that the United States had almost no lithium-ion industry. Virtually all the lithium-ion batteries sold in the United States were Asian. But Chamberlain wasn't bothered by the thought of American violators—he *wanted* the NMC to be used in the United States. He wanted Argonne to be noticed at home, and if any companies did decide to slip some of the NMC into their products—such as the new electrified cars under development by Detroit—he would easily sign them up to licenses. No American company would openly flout such a clear-cut, government-funded patent.

So he began talking to American companies that were in the battery game. He pushed them to shift to the NMC. Johnson Controls and Procter & Gamble both said they could in principle manufacture batteries installed with the NMC. But they would have to give it a long think. Configuring factories anew for a different battery would take five years. That was too slow for Chamberlain. He wanted Argonne technology inside American-made devices *now*.

That is when he recalled Khal Amine's electrolyte advance. Amine's additive required the alteration of no machinery. He returned to Johnson Controls and Procter & Gamble. Both companies agreed that an electrolyte modification would be simple. Chamberlain had his first American customers.

Chamberlain noticed other changes. For the first time, he was receiving eager calls of general interest from American companies. They had heard about the NMC and the electrolyte additive. What else did Chamberlain have on the shelf? "We started getting visitors," he said. Company managers wanted to see Argonne for themselves. The lab was becoming a player.

Chamberlain started to work up model licensing agreements

covering all of Argonne's battery portfolio. Companies had a choice: They could make an advance payment for rights to Amine's additive, the NMC, or any other invention, then pay regular fees under a ten-year licensing plan. Or they could opt for the equivalent of success fees based on sales. Chamberlain treated it all as part of his evolving, big-picture strategy. "Portfolio management is part of the economic war," he said.

12

A Little Talk with the South Koreans

Before long, word reached Chamberlain that the lithium-ion battery destined for GM's new Volt plug-in hybrid-electric—the car-making giant's attempt to take the early lead in electrified cars—would be supplied by LG Chemical, a South Korean company. Chamberlain also heard that the LG batteries relied on the NMC. He was not prepared to be seen as a sucker. He flew to Seoul.

Chamberlain was not going to sue LG—he doubted the U.S. government would approve a protracted patent battle, at least at the moment. For one thing, there was a strain of thinking that South Korea was not a direct competitor in the battery race. The argument was that the United States' best hope might be an alliance, and if that was the case, it was an easy decision with whom: China, with its own geostrategic ambitions, was an illogical choice; Japan had established a record of working alone and not sharing its advances. That left South Korea, which itself perceived the advantages of an arrangement with the United States. In 2010, the Obama administration singled out LG for a huge, $160 million slice of a $2.4 billion stimulus fund created to launch the United States into the global race to dominate advanced batteries. LG joined a core group of companies that, by building federally backed battery factories in the United States, would give American industry a reasonable hope of winning the race.

Republican critics assailed the fund. Their gripe wasn't what you might think—that the second-highest sum from the stimulus was going to a South Korean company. Instead, they complained that Obama was "picking winners." He was ordaining that batteries and electric cars would undergird the next great economic boom without any credible way of knowing that was the case.

They had a point—no one could be certain that the foreseen industries would materialize. But the counterfactuals swamped such misgivings, including that the race itself was indisputably under way—battery makers were working to improve lithium-ion, carmakers were designing and planning the launch of electric cars, and governments around the world were awarding grants and subsidies to support their given teams.

Around Argonne, there was particular affinity for the South Koreans. When South Korean scientists and companies toured the lab, they behaved so collaboratively that after some years, they were treated as honorary Americans. They simply did not seem to pose the same threat as the Chinese.

Still, Chamberlain himself believed that, while the United States aggressively pursued its place in the battery age, it also had to nail down what was rightly its own. That was why, even though the South Koreans might be commercial allies, he was now going after them. He wanted Argonne to be paid.

So, in Seoul, he informed LG that it was about to infringe on Argonne patents. When the South Koreans took the stimulus money from the Obama administration, they accepted fine print in which they pledged to respect American patents. Even though LG was not itself necessarily bound by the Argonne patent outside the United States, it was in effect inducing GM—an American company that *was* bound by the patent—to infringe, which would be a contractual violation. This astute bit of legalistic arm-twisting persuaded the South Koreans—LG had to license, and it did.

Now there was GM. Because LG already possessed a license to the NMC, GM was not technically obliged to pay for one as well. But the carmaker was intrigued with Chamberlain's talk of a higher-performance NMC 2.0. It specifically was captivated by work outside San Francisco by a small start-up called Envia

Systems that, collaborating with Argonne, had itself licensed the NMC and already begun carving out new ground for it. Envia had improved upon the capabilities of the Thackeray and Johnson cathode, an indicator that NMC 2.0 could soon break out of the lab and become a commercial reality. GM wanted to be the first to have it. The cost of its Volt hybrid electric cars could be slashed and its sales trajectory elevated. Moreover, it could also power a future pure electric car that, unlike the Volt, would have no gasoline-fueled backup engine. So GM paid for an NMC license. Attached to the agreement was a clause—if Argonne succeeded in creating the advanced version of the NMC, and GM could deploy it in its future electric cars, the lab would receive a substantial bonus payment.

Chamberlain, Thackeray, Johnson, and Amine were elated. The patent-free environment in which the Asian juggernaut was operating had been breached. These companies could not simply decide they were going to use Argonne technology with impunity. They had to pay for the intellectual property. That precedent captured the attention of nonprofit labs around the country, research centers that had been victim to the same offense, especially universities. Chamberlain felt he had arrived when an MIT intellectual property lawyer pulled him aside at a conference. "We wish *we* could cut these kind of deals," meaning the LG contract, the MIT man said. "With us, it's all about start-ups. But *you* license start-ups *and* the big companies. If it's a big company, you know the material is going to be manufactured."

Now Argonne was being noticed by the big guys. It was licensing not only to the garage outfits, the ones that might or might not even exist next year, but also to the GMs of the world, those that could actually use the licenses in products.

Chamberlain strolled through the Battery Department with numbers scribbled on mini Post-it notes. "You're going to have this added to your next paycheck," he told Chris Johnson, who, in protective glasses, was working at the bench. Johnson looked in disbelief.

"It's for the licenses," Chamberlain said.

The scientists were divvying up about $1.8 million. Chamberlain informed Thackeray, then Amine, and so on. Everywhere there was a stunned reaction. Chamberlain enjoyed it. As for the men, they held new respect for him. One day, Thackeray said, "You're a battery guy now."

In the summer of 2010, Argonne management assigned Chamberlain added duties—while retaining his IP licensing responsibilities, he was to lead the Battery Department. Thackeray intended no offense to Chamberlain's predecessor, but from a leadership standpoint, he thought the lab now changed for the better. He called Chamberlain "a real superstar." He might be expected to say that since it was his patent that was finally moving into automobiles, which is a big deal if that's been your life's work. But he meant to refer to Chamberlain's attention to the commercial cachet of the lab as a whole. The average government functionary was a wimp when it came to patent enforcement—federal labs were not commercial enterprises, after all. But not Chamberlain. As he left the lab in his $63,000 BMW one day, Amine said, "This car is bought from the licensing money." He credited Chamberlain. After a string of licensing deals, Xiaoping had new expectations. Amine would return home from the lab and she would ask, "Did you license again?"

The respect of peers is considerable juice for a scientist, but for Thackeray and Amine, commercialization was equivalent validation—and perhaps even greater. They would be eternally grateful to Chamberlain.

Thackeray saw the achievement from a highly personal perspective. In his thinking, these licenses meant that the "dominoes in the technology are beginning to fall." Science was shifting from the lab to the factory. Now, the sixty-two-year-old Thackeray was looking for NMC 2.0 to close out his career. "I just keep plugging away," he said. "I feel as though I got into a wave in South Africa and I am surfing. I am waiting to be dumped on a beach," but so far the latter had not happened. Thackeray felt good about his career. Chamberlain made him believe he could leave proudly once NMC 2.0 was out the door. "Now there are so many people, and very good people coming up," Thackeray said.

The prospects looked good for NMC 2.0—the United States was well positioned in the battery war—but Chamberlain knew that ultimately that was not good enough. It left electric cars as an iffy proposition—car buyers visiting a showroom would still not assess an electric car agnostically, not without a fairly large government subsidy. And even then, the electric car was likely to remain a "social purchase"—a product for the niche market of buyers wishing association with the green movement. The battery guys had further to go.

13

What Andy Grove Said

I n October 2009, Andrew Grove, the former chairman of Intel and a father of Silicon Valley, asked Chamberlain to bring his Argonne team to his house in California. When they were before him, Grove surveyed the battery race and said that when he looked at lithium-ion, he saw silicon chips. The American chip industry blossomed in the 1950s when the first integrated circuits were made on microchips. By the 1980s, though, Japan had swallowed up much of what by then had evolved into a commodity business, along with its jobs. Grove managed to salvage Intel and give it a renaissance by rejiggering the equation—he transformed it into a microprocessor company, producing computers on a chip, and not just the chips themselves. With batteries, he said, there was also a brutal, dual competition under way—for who would finally make them more powerful, last longer, and be safer, and who would manufacture the discovery once it was made.

But Grove thought that the stakes were far higher than when the Japanese were capturing the chip market. He saw a "clear and present danger" that if the United States could not account for its own energy requirements, it would later be at the political and economic mercy of those who could. The United States had to create a battery-driven fleet of private vehicles to replace much of the current gasoline-powered ones. As passionate as he had ever been about anything, he had telephoned his successors at Intel and urged them to take on this new fight—to get into the advanced battery business. They more or less brushed him off—for one thing, unlike the fixed physics of silicon, batteries were based on electrochemistry, which changed constantly. Batteries were too complicated. Which is what led Grove to Chamberlain.

The Argonne scientists arrived with a three-phase plan to test on Grove, who perhaps could do for them what he had accomplished for Silicon Valley. To make it easy to digest, they compressed it into bullet points on a single page: Current lithium-ion—the NMC and what was going into the first new electric cars—was phase one. Phase two was the next three to five years, in which they would push out "advanced lithium-ion," meaning NMC 2.0, a bridging technology that would possess twice to three times the longevity of current lithium-ion batteries. That would be sufficient to carry hybrid and electric cars for a decade or two, a space of time in which to develop phase three—"the brass ring," as Chamberlain called it. This final phase involved a technology named lithium-air. It was ultimately where Argonne would go.

Chamberlain imagined that Wan Gang and his team were strategizing similarly. Walking the Argonne men to the door, Grove urged them to remember one thing: "Winning a manufacturing game is nine rounds. It's a boxing match. You will lose the first round, and probably the first two rounds. But always remember—the game is nine rounds."

14

How to Navigate Great Minds

Thackeray and Amine could work anywhere. They could leave Argonne and strike out on their own, for instance. If they did so, one advantage would be leaving behind the requirement to vet their project ideas through Department of Energy bureaucrats, who could demand many pages of documentation solely to explain whether a new experiment would be safe. Yet they wouldn't necessarily earn more money working in the private sector—government rules established in 1980 allowed recipients of federal grants, such as government scientists and university professors, to retain total or partial rights to their patents. Both scientists were reaping the rewards of their NMC work.

There were other pluses to working in a national lab, factors of which Chamberlain kept Thackeray and Amine aware as part of his job to manage sensitive souls with enormous egos. First and foremost was that, if you wanted to publish in respected journals and have your patents noticed, Argonne was a good home. The battery scientists were competitive—everyone knew they were in one of the hottest fields in science or technology and all wanted to make the next advance first and have it quickly adopted commercially. Argonne was on the leading edge of this particular science. The battery guys were sheepish about saying so explicitly, but they felt they fielded the strongest collective team in the country and perhaps the world, retaining their stars and attracting smart, fresh minds as postdoctoral assistants. If someone on the outside suddenly edged ahead, Chamberlain's team became agitated.

It aggravated Thackeray to learn that Envia, the California start-up, managed to make batteries work 50 percent longer than his own team could. Thackeray's wholly predictable reaction allowed Chamberlain to do what he did best, what he called "managing from the backseat." Even though Envia was effectively an appendage of Argonne—it was improving on the lab's own patent—Thackeray regarded it as *competition*. He began to push his team harder.

The battery guys would tell you that Chamberlain was hard to get in to see—he was so busy with his double job—but that when you did sit down with him, he tried to understand your ambitions and to create the opportunity for them to be realized. They cherished him for that quality. But they also knew that Chamberlain was a calculating boss with fixed aims, and the most fixed of all was the race. In practice, one meaning was that, even though Argonne was a government lab, technically owned by the public, they could not publish everything they learned for the very reason that a race was under way. It was an iterative game. You wanted everyone to publish so that the field as a whole moved forward—and the scientists earned the peer respect accorded to prolific writing. But you also hoped to protect your own inventions—as soon as you published, you enabled your competition to use your inventions against you. Thackeray and Amine—everyone—understood that calculus. It was part of Chamberlain's war-within-a-war.

Chamberlain drove eighteen miles from home to Argonne every day along 75th Street through the suburbs that had grown up in the decades since the lab first plopped down in Tulgey Wood. He turned on Cass Avenue South, which eventually became a wide, picturesque road flanked by forest. Before long, he turned sharply right at a big, black sign containing Argonne's logo—a triangle whose flanks were painted the primary colors. A mile-long drive followed through the forest planted at the beginning and ended at a guard shack where Chamberlain would show the badge hanging around his neck. Then he turned right into a sea of mostly red brick buildings, spread over the hundreds of acres. Within minutes, he would see it—Building 205, on a smaller black sign, also embossed with the logo. After fifty or so footsteps, he would be in the building.

Today, Chamberlain made a left toward the "dry room." It was a state-of-the-art, moisture-proof setting customized for fiddling with advanced lithium-ion batteries. An air lock separated the dry room from the outside world. Inside, a slurry of carbon and NMC was coated onto rolls of aluminum, creating new battery cathodes. "It's a nineteenth-century technology," Chamberlain said, deserving of no place in such a lab. In labs he had seen in other countries, Chamberlain added in a whisper, scientists actually stood by and dipped a finger into the slurry in order to pass judgment on its quality. When Wan Gang was at Argonne, he was only Chamberlain's latest Chinese visitor. A few months previous, the Americans raised the subject of lithium-air with a Chinese delegation and offered to help them develop the technology. But the visitors wanted to discuss only one subject—lithium-ion. The theme was: how do we get what you invented? Meaning the NMC. Chamberlain and his team were extremely cautious, silent. Argonne could build robots that automatically created a perfect slurry. It could invent the NMC. But China could not. "That's where we will catch up," Chamberlain said.

Perhaps Chamberlain was right. Thackeray and Amine could be the ones to make it happen.

PART II

FOREIGNERS IN THE LAB

15

The Start-up

I n June 2007, Sujeet Kumar and Mike Sinkula took up office at the public library in downtown Palo Alto. Kumar, a battery scientist born in India, and Sinkula, a San Francisco native, were using a public conference room there to talk on cell phones, print and fax papers, and organize meetings. For four months, they worked to raise $3.2 million in order to license and validate the idea that Argonne's NMC 2.0, the advanced cathode that resulted from jolting NMC with a little more than 4.5 volts of electricity, could double the capacity and halve the cost of automotive batteries. They would spend the money to form a team of ten or so scientists who would build such an ultrapowerful battery. That would allow them to raise more cash and aim to manufacture it.

At the time, it appeared that, in the automotive world, only Kumar and Sinkula were aware of Argonne's NMC formulation. They alone seemed to conclude that, if you were thinking of profit-making applications, NMC 2.0 was the most promising battery material available. Kumar had identified its properties at the Silicon Valley start-up company where he and Sinkula previously worked. The start-up, NanoeXa, sought to develop batteries for power tools, and Kumar, its chief engineer, had stumbled on the NMC after a months-long hunt for an edge over incumbent companies. He had examined hundreds of patents and academic papers on lithium-ion. The NMC and its second-generation improvement were superior to anything else he found. In 2006, he surprised

Thackeray with a call about the invention. Until then, Thackeray had detected no industry interest in it at all, but NanoeXa's CEO proceeded to license it for $150,000.

Just a few months later, Kumar and Sinkula both left NanoeXa, explaining that they wanted to work "in another company."[1] To reassure their surprised CEO, they said it would not be associated with the NMC market. Within a few weeks, though, Argonne received another call from Kumar—he and Sinkula had launched their own start-up company. It would center around the NMC and be marketed to carmakers. In the coming years, the move on their own would be the subject of a considerable dispute with Michael Pak, NanoeXa's CEO. But for now, fortune was with them. As Jeff Chamberlain had found in his own start-up stage, energy was the rage in Silicon Valley. Venture capital firms were competing fiercely for the most promising ideas. They had decided that renewable energy was the next big boom. But their eagerness seemed different from the past manias. It wasn't just about money. The fever aligned with the Valley's strain of politics, which generally vilified oil, embraced its technological rivals, and fretted about climate change. Here was a way for the venture capitalists to do well and do good.

Nationally and globally, a similar sentiment took hold about global warming. Barack Obama, at the time an American senator initiating a campaign for president, vowed to promote non–fossil fuel technology and reduce emissions of heat-trapping gases. But it was generally believed that whoever was elected, Democrat or Republican, would push through laws and federal spending to buoy solar, wind, biofuel—and battery companies. Silicon Valley's venture capital community was prepared for these new policies and the commerce that would follow.

So Kumar sensed an almost physical reaction when, sitting before VCs, he launched his PowerPoint slide deck and said that the NMC, although more advanced than anything on the market, had thus far gone unnoticed. In Boston, a Harvard Business School graduate at a firm called RockPort Capital Partners "got very excited," Kumar said. He asked to see Thackeray or Amine for himself. If what Kumar and Sinkula said was true, they would have their $3.2 million.

A couple of weeks later, Kumar, Sinkula, and two RockPort men flew to Chicago. Hearing out Amine, the investors returned to Boston and e-mailed a "term sheet," an official commitment to fund them in exchange for half of their as-yet-unnamed company, to Kumar and Sinkula.

This was fast for Kumar. Despite his experience in start-ups, he now realized that he did not actually understand how funding worked. Back home in Fremont, California, he telephoned a college classmate of his wife's.

Atul Kapadia was a principal at Bay Partners, one of the Valley's oldest venture capital firms.

"RockPort is funding me, but I don't understand the term sheet," Kumar said. "Can you help?"

Mumbai-born Kapadia had an MBA from Stanford and a bachelor's in biomedical engineering from the University of Bombay. He had gone to work at Sun Microsystems straight out of school, assigned to a design team that produced the Spitfire, a chip that later would be called the fastest in the world.

Kapadia had questions, starting with why Kumar was seeking money on the East Coast when the biggest VC firms were in the Valley. He suggested that Kumar and Sinkula drop by his Palo Alto office with their slide deck.

Two hours later, after hearing out the pair, Kapadia rang Kumar's cell: *I* will fund the idea, he said.

Kapadia offered essentially the same terms as the Boston group—$3.2 million for half the start-up, a standard offer in Silicon Valley venture capital. But as a demonstration of goodwill, he was prepared to cut Kumar a check immediately for half a million dollars so he could start working right away. Kumar could consider it a loan until the details of the equity investment were arranged.

Looking back later, Kumar had the feeling that he could have raised much more—$5 million or even $6 million—for the same equity share. His idea clearly was solid—after just a few calls, he and Sinkula already had two bites. But that was hindsight. Now he had to create a company. He accepted Kapadia's rival offer.

He and Sinkula called their company "Envia Systems" and opened their lab in a small, new industrial park in an East Bay suburb called Newark, tucked amid a hive of tech start-ups. The company name was Sinkula's idea, combining "en" from the word "energy" and "via," from the Spanish for "way"—the way to energy.

Kumar called Chamberlain. "I'm ready to license again."

Kumar's agility surprised Chamberlain. He had orchestrated the deal based almost solely on an idea: he had no prototype in hand. As for intellectual property, he had only a written offer from Chamberlain for a nonexclusive license. "How did he pull it off when these VC guys are so cutthroat?" Chamberlain wondered. One lesson again was the timing—Kumar and Sinkula had struck precisely when the market demanded someone like them. But there was more to it. Venture capitalists did not say so explicitly, but technology, while important, was not their primary concern. They looked closely at a venture's management. Their key question was, "Have they done this before?" Meaning, had the entrepreneurs involved carried out a prior venture deal that resulted in a big return? If they had, the investors would be exceptionally courteous and forthcoming with their cash. Chamberlain acted out the repartee in such a case:

> You have already made my fund hundreds of millions
> of dollars. You are here to ask permission to do it again
> for me? Yes, please. How much money do you need?

Kumar had not earned anyone hundreds of millions of dollars, but he had "done it"—he had helped to lead three previous start-ups, which again was the core credential. It was what made the difference. He could approach VCs and say, "Look, here is a new opportunity. I don't have the license deal yet, but you have to act fast. Here are the terms we've agreed on with Argonne." "And, bang," Chamberlain said, "you have three million dollars. He got it immediately."

By comparison, Chamberlain had crawled Sand Hill Road just a couple of years before. When asked about his prior experience, Chamberlain replied that he "absolutely" had built a business from scratch—at Cabot. But venture capitalists felt differently about achievements within a big corporate setting. It wasn't the same as success on a tight budget with a small staff and the risk on your own shoulders. Their reaction was "No, you haven't done it."

At Argonne, Chamberlain stood out because of his industry credentials and the lucrative deals he had negotiated. But a start-up was different. Kumar showed how a star played the venture capital game.

16

Out of India (and China and Africa)

Kumar greeted you in a high, halting lilt, his black hair flopped flat to the side. He wore plain, long-sleeve dress shirts and conservative slacks and evaluated you through rectangular specs. The picture was dated, rumpled, and bumpkinesque. This might have been a problem given a Silicon Valley CEO's need to impress a variety of sophisticated outsiders. But then he would talk batteries and unpretentious authenticity would surface. Meeting him face to face at a Fort Lauderdale battery conference, Chamberlain sensed "a man of his word. He would do what he said." Sinkula—with his stylishly cut hair, slacks, and jacket—was clearly the businessman of the two. Sizing them up, Chamberlain saw "two young, aggressive entrepreneurs."

The eastern Indian city of Patna, where Kumar grew up, is situated along the south bank of the Ganges River. After school, Kumar and his friends played cricket and football in vacant lots that dotted their neighborhood. At dusk, the electric grid would be overwhelmed for three or four hours, sending the homes into darkness, and he would study by kerosene lamp before scurrying back outside for more play in the dark. He was the excessively pampered youngest of six children, and his father finally decided that this circumstance was hampering his education. At ten, Kumar was packed off to board at a military school. "I cried a lot—every time I had to leave, I cried," Kumar said. But by the time he arrived at school in Hazaribagh, a forested, mountainous area five hours south of Patna, he would tumble into play with the rest of the ten-year-olds, the sorrow forgotten.

Each year, some 100,000 Indian students completed the entrance exam for 2,000 slots in India's exclusive Institutes of Technology (IIT), a 2 percent admission rate. By comparison, Harvard University admitted *16 percent* of its applicants in 1986, the year Kumar had to decide his college future. But there was substantial motivation for the multitudes who tried—those admitted to one of the Institute's seven campuses went on to make up the cream of Indian science and technology, on par with the best in America, or anywhere. Graduates in fact often moved to Silicon Valley and assumed dominant roles there. Those who took the test often carried with them the lifelong ambitions of their entire extended family. If you were successful, Kumar said, "your life was set."

Kumar aspired to attend IIT. And when he took the exam, he scored in the top 1,800, winning him a place in the Institutes' statistics program, a cause for jubilation. But in his case it wasn't. He aspired not only to study at the Institutes, but admission to their engineering program. *Those* were the IIT grads who were admired—they were whom you pictured when hearing of an alumnus; barely anyone knew that IIT even offered statistics. But Kumar needed a much better score to qualify for engineering. He felt more dejected than thrilled.

One of his brothers said he was misinformed—*admission* to the Institutes was the point, whether in statistics or any other program, and he ought to "get into IIT any way I could." But Kumar's father echoed his own thinking. "Why do you want to study statistics?" he said. "You should go into engineering."

There was another option. Kumar could use his still very good test score to study at the technological institute in Varanasi, a four-hour drive from Patna, almost straight west along the Ganges. Varanasi was a distinguished school—not long after, it in fact would be absorbed by the Institutes. But its graduates typically went on to uncreative jobs in steel or other basic industries, which to Kumar was a numbing thought. He didn't know what would be worse—to study statistics or work in the steel industry. His parents stepped in: Varanasi was relatively close by, which comforted his mother; engineering was the more secure career, which decided his father. The teen packed for Varanasi.

Three years into his studies, stories began to trickle in from the United States. Recent Varanasi alumni spoke of scholarships for doctoral programs and good, high-paying jobs for top graduates. Institutes of Technology alumni had the pick of the tech and engineering jobs at home, but in the United States your school was not necessarily the main thing. As long as you were talented, you could be on the same track as IIT grads.

Kumar opted to try. In his senior year, he studied for the Graduate Record Examinations (GREs) and scoured descriptions of American universities. A letter from the United States arrived. The University of Rochester offered a full scholarship including a thousand-dollar-a-month stipend. He would have to work as a teaching assistant and in other campus jobs. But he would have a full ride toward a Ph.D.

No one in the Kumar family had ever studied abroad. His mother had discouraged him from even trying. But now that he was in, there was no question that he should go. His father found the money for his son's airfare. Kumar landed in New York all but penniless. He was twenty-one.

Kumar arrived on the cusp of a powerful surge of Asian immigrants into Silicon Valley and throughout the American technology industry. Asians already held a third of the technology jobs in the Valley and were half of the software developers.[1] More than a quarter of American doctoral degree recipients as a whole were foreign born and half of those from Asia.[2] In engineering and computer science, about 40 percent of the Ph.D.s were born abroad.[3]

These demographics created racial tension. A black leader in Silicon Valley said the Asian employment bulge was not incidental—technology companies, she said, "do not want to employ Americans.

They import labor from overseas, pushing for H-1B visas," residency permits allotted annually to specially skilled foreigners. It was true that there were relatively few blacks in Silicon Valley technology. But there was no evidence that any particular racial group was favored or excluded. In the case of battery guys in particular, other dynamics seemed at play in the appearance of partiality.

When you looked around a quarter century later, one of the first things you noticed in the battery race was the trend's deep roots—America's battery team was largely foreign born. There was the occasional American-born battery guy—the families of most of the researchers on Thackeray's small team had been in the United States for generations, as had Chamberlain's. But Thackeray himself was born in Pretoria. Chamberlain's deputy, Tony Burrell, was from Palmerston North, on New Zealand's North Island. Chamberlain's immediate boss, Emilio Bunel, was Chilean.

The same was true across the American battery brain trust: though John Goodenough grew up in Connecticut, Stanford's Yi Cui was born in China, Berkeley's Venkat Srinivasan in India, and MIT's Yet-Ming Chiang in Taiwan. In the industry, not just Sujeet Kumar and Atul Kapadia but almost their entire team of scientists was born in India.

Moroccan-born Khalil Amine unapologetically hired only foreigners. His group included not a single American-born researcher. Over the years, Amine had employed the occasional American and even a Frenchman. But now, apart from two other Moroccans (and himself), his group was entirely Chinese. Over sushi after work, Amine said he had concluded that the job was too demanding for United States–born Americans. And not just for them—some Asians, too, were not up to the task. "I have had Caucasians in my group before. Also Indians, Koreans," Amine said. "But I will tell you this— I'm very demanding. I come to work at six A.M., five A.M. I work weekends. I have to make sure that we produce. The Chinese work this way, too—they are extremely hardworking. But some of the Caucasians, they don't like that. It seems like big stress on them."

Amine was not alone in invoking a supposedly unique Asian

cultural DNA when it came to science, technology, and the work ethic, in particular one native to Chinese, but he said the results spoke for themselves. If you considered inventions and published papers, his group was the most prolific in the Battery Department. By Amine's own count, his group had produced 120 or so inventions over the last decade. "The next group is not even close," he said, which was true. "And if you look at papers—last year we published about forty-seven, forty-eight. Some professors, they publish that many in their entire careers." Amine himself had been awarded thirty-eight patents since arriving at Argonne. The next-highest recipient in the department—Thackeray—had twenty-four. Numbers alone were not a definitive metric—China turned out a torrent of forgettable patents and papers. But American patents were not as easily obtained. For Amine, they told the story of his group's stature.

The subtext wasn't merely the view that foreign-born battery guys worked harder but that Americans were simply not a large part of the job pool. The battery guys said that when they advertised a new position, dozens of applicants would respond of whom just two or three typically would be American. The proportions explained why these few Americans, whatever their qualifications, were often outshined by the mountain of overseas competition.

There simply did not seem to be many Americans eager to invent the next big battery. Americans trained in the disciplines attacking the battery challenge—in physics, chemical engineering, material science. But their jobs of choice tended to be in other fields. Among the places they landed were Silicon Valley's high-tech firms. Or, even if they did go into batteries, they rejected basic research, which almost certainly required up to three years of uncertain toil as a postdoctoral assistant, and went into private industry.

The story was similar with most foreign students. They piled into information technology jobs in the 1990s. High-tech industry demand for them as engineers and other specialists was substantial, and although the necessary H-1B visas were limited, computer hardware and software companies often managed to obtain them for those it sought. One could live very well and even become rich in such jobs.

But there was also a large well of foreign students attracted to batteries. Perhaps again it was a simple matter of proportions—when there were so many foreign Ph.D.s, it stood to reason that would-be battery guys would make up a certain minority. But one trait of Argonne's foreign-born staff was traditional personal and family aspirations: they were seeking a new life with greater prospects for their children. "I'm not saying it in a way to degrade the other guys," Amine said, "but Caucasian Americans—they don't want to do Ph.D.s. They go for an MBA or something like that. For example, I was invited to give a talk at MIT. I would say seventy percent of the students were Asian. Chinese, Koreans, and Japanese. I went to Berkeley— same thing." Foreign battery guys in fact often completed not just one postdoctoral assistantship before securing permanent employment, but two or even three three-year stints. A postdoctoral researcher at Argonne earned about $61,000 a year, which was high for such a position. When offered a staff job, the pay was bumped up a bit and rose regularly from there, which became even more attractive in combination with the stability of federal lab work. But it was not high-tech scale. Their determination was distinct not just from Americans' but also from that of the Silicon Valley immigrants.

Once you settled on a life in batteries, a simple calculus made Argonne and the other national labs special magnets for such foreign Ph.D.s—the number of private battery companies was small and with it the possibility of obtaining an H-1B visa. The national labs, on the other hand, could sponsor an unlimited number of H-1Bs—in 2000, Congress had created a working visa exemption for nonprofit, university, and national labs.

The national labs served as a miniature Lower East Side—not quite a teeming melting pot, but a greenhouse where battery-minded immigrants were invited to succeed and, if they did and so desired, could take citizenship. "They go an extra length. They're smart. And they are extremely reliable," Amine said. Why was his team predominantly Chinese? "That's why," he said.

Amine said his strategy did not always work in his favor. He had lost numerous military contracts because the Pentagon permitted only American citizens to work on such sensitive projects, and his group lacked them. But he was straightening that out, too. Six years earlier, Amine himself had taken American citizenship. His two Moroccan researchers had as well, and a Chinese scientist was on his way. "I think within five years, all these Chinese will be U.S. citizens," Amine said. "It's just a matter of time."

Ultimately, Amine said, his personnel preferences were unimportant. "At Argonne, the policy is you hire people based on capability. Not nationality," he said. Of course, Amine had determined that there was a difference—he *was* hiring according to nationality. It was among the reasons why an American victory in the battery race oddly depended on scientists from rival countries.

Kumar finished his Ph.D. in 1996. The most obvious places to apply for work were Kodak and Xerox, both of which had large local research labs. But neither company was hiring. Then Kumar noticed a job posted at a university—a California professor was looking for a postdoctoral assistant to help him set up a battery company. Until then, Kumar had not considered electrochemistry as a profession. Nor did he have any expertise in batteries. But the professor was persuasive—Kumar would be his first employee, he said. The professor was confident the start-up, called Nanogram, would succeed, making the early employees wealthy. They would use nanotechnology to create batteries for medical devices. Kumar would be responsible for setting up the entire lab. He took the job.

Eight years later, Nanogram was sold, resulting in a $300,000 payout to Kumar for the shares he held, a sum that gave him financial stability for the first time in his life. He put it away in the bank. Two years after that, Kumar was hired as director of technology for NanoeXa, the battery company that first licensed Argonne's NMC.

NanoeXa's CEO, Michael Pak, was not a battery guy. He was a

South Korean–born businessman with a Harvard undergraduate degree and a knack for attracting investment cash in Silicon Valley. He had raised money for start-ups creating video games and fuel cells. Now, Pak was interested in lithium-ion batteries and he relied on Kumar to guide the company to the right technology.

When they licensed the Argonne material, Kumar's objective was to use it for power tools. He tinkered with the NMC for some months and then applied for a patent for a NanoeXa version of the cathode. But Pak saw the license additionally as a valuable bargaining chip. He went with a proposal to a South Korean flat panel display company called Decktron: he would exchange his ultra-valuable Argonne license—exclusive global rights to six of the lab's patents in all, he said, worth $21 million in commercial terms—for majority control of Decktron. In late 2006, the deal was consummated. Pak and Kumar took seats on the Decktron board, with Pak as CEO and president.

The problem was that the Argonne license *wasn't* exclusive. It allowed Pak to manufacture the NMC for power tools, but Argonne was free to license its patents to anyone else it pleased. Moreover, Pak was not permitted to bargain away the licenses as currency to buy into other companies. Two years later, Decktron collapsed and was delisted from the Korean exchange.

But by then, Kumar—comforted by his nest egg in the bank—and Sinkula had moved on. Along the way, Kumar had found he possessed an unusual instinct for the atomic-level physics of batteries.

17

Why We Stay in Chicago

Just as a former professor prevailed upon Wan Gang to leave Germany and help to create the future at home, one might think that Chinese battery guys in the United States would also return. It made sense that they would feel the pull of the homeland. And if these impulses did draw a certain number of key players home to China, and other immigrants to their countries, the American team would be hollowed out.

But that was not what was happening. Government incentives were attracting increasing numbers of Chinese students to repatriate but this trend largely excluded the staff at Argonne. Of the lab's foreign researchers, the Chinese were among the least likely to repatriate. Two South Koreans returned home after some years at Argonne, persuaded in part by the excellent public education system that Korea offered their children, but Chinese and Indian scientists did not seem to even seriously contemplate going back. The professional conditions in China were a disincentive. Unless you were an established name player, the way that Wan had been in Germany, you could end up lost in a sprawling lab in your native country, serving an autocratic boss interested not in new ideas but largely in retaining his own position. Whereas at Argonne your ideas, as long as they were competitive, stood a reasonable chance of being funded.

Zonghai Chen, a thirty-eight-year-old researcher from a tea-growing town called Penglai in Fujian province, said Internet ads from Chinese laboratories promised one-million-yuan salaries, equivalent to $160,000 a year, 50 percent more than he currently earned. The new positions for mid-career scientists aimed to fill a hole in the job pool created by the Cultural Revolution, when a generation of scientists went missing. "The old researchers retired, and the young ones are not ready to take up their posts," Zonghai said. "So they are offering big packages to go back."

Zonghai said he had not responded to the ads. The money was alluring, but after more than a decade in the West, a period in which both of his children were born—a six-year-old girl and a nine-year-old boy—higher pay alone was not sufficient to leave Argonne. "My children would have to learn Chinese as a foreign language," Zonghai said. "If my daughter started now, she would be on a reading level lower than the first grade. She would have to catch up fast."

Several of the Chinese staff cited this language gap, a nightmare for Americanized children who would face competition with classmates already well along in their education. Their children might simply be unable to compete, a gamble they were not prepared to take. Yang Ren, a scientist on Argonne's Advanced Photon Source, a gigantic X-ray loop that researchers used to examine their creations, said that was not all—it would frankly be a step down to work in a Chinese setting. "If you want to do good science, it is here," said the forty-eight-year-old Yang, who was from Anhui province.

Amine's largely Chinese staff concerned some of the other managers. Theirs was a common story of American immigrants: Argonne managers said the Chinese kept to themselves and that no one knew what they were doing or thinking. During a walk back from the cafeteria, an American postdoctoral assistant said Khalil Amine's staff members were very, very good at their work, "but," he added, "I am wondering about the national implications." Amine's calculation was, "How do I get ten papers out this year?" The American went on, "It is the easiest to hire Chinese postdocs and achieve that. But what about the long-term, national strategic mission? Basically, are we training the athletes on the other side?"

The reactions to Amine's hiring and working methods were not nonsensical. One needed only to recall his assertion that Thackeray stole his idea for the NMC patent, when he shut off his staff from everyone else in the lab. But there were reasons for Amine's behavior. If you wanted to understand him, you had to consider his experience in Japan. In the United States, Amine's observation was that innovations often went nowhere. A university professor would make a discovery, apply for a patent, and move on to the next idea. In most cases, the patent would remain unnoticed. If a company happened to unearth it, the university would gladly sign it over with exclusive rights for a few hundred thousand dollars. Two years later, Amine said, you would notice the product for sale, with the company making the money. From the standpoint of just rewards, that could discourage an inventor the next time a bright idea came to mind. It could seem almost worse, he said, to see someone else profiting from your hard work than not to have it commercialized at all.

Japanese universities were mindful of the scarcity of exceptional ideas. When corporations came calling, the universities tended to be harder bargainers than their American counterparts.

Argonne employed some three thousand scientists but Amine was appalled at its relatively small intellectual property unit. The lab seemed content to file away strong inventions without seeking publicity. There was no explaining it apart from either a diffidence toward the business of science or plain languor. Whichever, Argonne's IP team was passive when it came to licensing the lab's inventions. So Amine set out to create his own little Japan. Amine organized his staff along the lines of the Kyoto invention machine where he learned his craft. He whipped his researchers into a cadre that at his direction worked systematically through every possible approach to the solution of a chemical puzzle—hundreds if necessary. The enviable record of papers, patents, and industry interest followed.

Of one of his Chinese researchers, Amine said, "When you give him an experiment, he does it fast. He'll give you the result in two days. With some people it's like pulling teeth."

Amine's critics pilloried his record of picking up a promising idea produced elsewhere, blending it with his own flashes of intuition and the work of his efficient staff, and emerging with a patent application or a new paper. They insinuated that it was theft. But in Japan—or any of the big Asian manufacturing economies—his methods would be recognized as fair and even sensible. Japan, China, and South Korea continued to retain their economic edge with a willingness to build on others' ideas and spend money for years and years with the confidence that a profitable industry would eventually result. Amine was merely following the Japanese way.

As critical as they were of him, Amine was savage toward the usual practices in American industry and labs. Western scientists championed the visionary moment but that led to "the moon or nothing. So they have nothing," he said. He was prepared to go step by step. And he winnowed down his group to those who would work the way he saw fit. That meant only two nationalities—Chinese and Moroccans.

On its face, Amine's hiring sounded racist. His management style was dictatorial. But Amine was neither unethical nor a bigot. Rather, he was opportunistic in noticing others' advances, uncanny in identifying and resolving a flaw, and ruthless in cutting through to a product bearing his name. That made him no different from countless other successful Americans.

Jun Lu, a researcher on futuristic lithium-air batteries, defended Amine's Japanese notions. Jun and his wife, Temping Yu, who also worked at Argonne, had no relatives in the Chicago area. "So we have more time to focus on research. You work harder" on Amine's team, he said, but that was only part of the picture. "If you want to be successful, you still have to have the ideas. You have to have common sense."

But there were also pockets of anger in Amine's group. This

was not Japan. Some members of his group did not appreciate serving as cogs in Amine's machine rather than innovators and thinkers in their own right. Amine held out the coin of the realm—an American visa and the later hope of citizenship. Their names appeared on the papers to which their grunt work contributed. But some of Amine's best staff bristled at his regimentation, seeing the arrangement as indentured servitude. Two of Amine's most talented scientists—both of them principal researchers on the NMC—returned to South Korea after painful experiences under him. Amine regarded himself as a keen judge of talent—he thought he knew who was who and how to incentivize them. But his high-throughput approach didn't always work out.

There was a divide between the Chinese and the rest of the battery department. The Americans were suspicious of the Chinese and also themselves insular. The old days of Argonne scientists hanging out at one another's homes were long past—in 2011, five years after he joined the lab, Chamberlain had yet to throw a party. Almost none of the battery guys had ever been to his house. An administrative staff member's ears perked up when her boss mentioned dinner plans with a colleague—it was the first time she had ever heard of lab executives socializing together. She could only speculate why so little entertaining went on. It wasn't that the scientists were unfriendly. But there seemed to be an unspoken midwestern distance. Andy Jansen and Kevin Gallagher, both battery guys, threw backyard barbecues for department colleagues, but Asians were rarely present. Once, when Gallagher, a young engineer, brought along a South Korean scientist named Sun-Ho Kang to a gathering, Janssen asked why he had not joined them before.

"I was never invited," Kang said.

As a young man living in Seoul, Kang had wanted to see the world. His father, a construction subcontractor who supported the family by renting out the couple of forklifts that he owned, had never left South Korea. Neither had his sister, Kang Eun Kyung, though she was a well-known composer of pop lyrics. A professor recommended that Kang contact a physicist he knew—John Goodenough, then a professor at the University of Texas at Austin. Another of his students had been a postdoctoral assistant under the battery pioneer, but he was moving on, so there might be an open slot in Goodenough's lab.

Goodenough responded quickly when Kang e-mailed—the position was his as soon as he wished to show up. "Somehow he had the impression that South Korean students—that anyone from my university—should be good," Kang said.

When Kang arrived with his wife and daughter, they found Austin expensive—his salary was two thousand dollars a month, leaving little after eight hundred dollars in rent. But working with Goodenough was itself compensation. Kang found his new mentor to be "just a genius."

"What do you want to do?" Goodenough asked when they met. Kang had never researched lithium-ion, so Goodenough gave him an assignment: a visiting scientist was submitting a paper to *Nature*, but Goodenough was suspicious of his lab results.

"Try to reproduce them," he said.

When Kang reported his findings, Goodenough concluded that the claims were in fact faked. The visiting scientist was attempting to snooker *Nature*. Such work could not come out of Goodenough's lab.

"And that was the start of my luck," Kang said. Over the next year and a half, he and Goodenough worked on cathodes and super-capacitors, energy storage devices that deliver a burst of power for a short period. "Having my name next to his on a paper was an honor for me," Kang said.

One day, Kang noticed a circular on Goodenough's desk—the announcement of a job opening at Argonne. For the first time, Kang confided his financial circumstances.

"I don't have funding to raise your salary," the professor said.

"I understand, so I'd like to apply for that position," Kang said. Goodenough handed him the announcement.

Don Vissers, the same manager who had recruited Thackeray and Amine, responded warmly. Kang moved to Chicago with a position on Khalil Amine's team at double his Austin pay.

It was not long before Kang felt like "a workhorse." He was carrying out repetitive tasks in which Amine was attempting again and again to advance yet another theory that would produce yet another paper or patent "that doesn't change anything." The Moroccan traveled frequently but provided his subordinates no opportunity to attend the same international conferences, mix with peers, or make a name for *themselves*. None of his staffers won promotion for their work. Kang imagined that, should he carry on, he would retire as he was—a research scientist. Kang shrank into himself—he "tried to be a nobody." "Maybe that sounds weird. But that was my attitude," he said. "I didn't expect anything from the group."

The then-department head transferred Kang to another manager. His best work followed, including a crucial role in the development of NMC 2.0. Under Thackeray, Kang discovered inner truths about the material that no one else recognized. He at last found close friends on the staff, in particular Gallagher, the young engineer. But it was too late. Kang wanted to "contribute to everyday life," to work in applied science. Argonne seemed too far removed.

About this time, he attended a Chicago dinner hosted by Park Sang-Jin, the CEO of Samsung SDI, the battery division of the South Korean conglomerate. One dinner topic was innovation. Kang said that, at Argonne, he did not try "to do everything alone. I know it is much faster and more effective when I find someone and try to collaborate." Park said the same—"nobody can do it all."

The meeting was important for Park. Samsung had eclipsed Japanese companies and become the world's largest maker of lithium-ion batteries for electronics. At that moment, the company was aiming to expand and develop advanced automobile batteries as well, challenging former juggernaut Japan along the entire commercial chain, from energy storage to consumer products. As the incumbent, Japan was still ahead in the battery race—its laboratories were superior, it had a decade-and-a-half head start on the factory floor, and its brand names were still prized. But Samsung was capturing a greater share of the total market. In the case of vehicular batteries, the trouble was that Samsung lacked expertise with the very different cathodes required for the ten-year life demanded of them. With his collaborative part in the work on NMC 2.0, Kang seemed ideal to lead this research. Park asked Kang to join the company. His title would be vice president. That suited Kang, because "I didn't want to go somewhere and be some very minor person."

He and his family prepared to return to Seoul.

Considering where he was at Argonne, the scale of the elevation was stunning. He looked back. Eleven years earlier, he had left South Korea to experience the world. At Argonne, he had anguished over a decidedly subordinate role in Amine's research group. Now he was returning home as a leading member of South Korea's team in the battery race. "They expect me to solve the problem," Kang said. He would "taste industry and challenge myself—will I survive or not?"

Even when he was working on NMC 2.0, Kang thought he had not been particularly creative. "I wished, I wished," Kang said. "But I just followed. All I did was try to be open." Kang said that to be innovative, "people should first be very desperate. Otherwise they don't need innovation." That was why the Chinese would lose the battery war—they typically moved slowly when it came to technology because they did not have to do otherwise. Historically, they borrowed technology from others.

Americans, Kang said, had more potential than almost anyone because they had the fundamentals—from childhood, they were trained to argue and discuss. But they, too, were handicapped: they were not desperate. "They are not prepared to lose everything." At Argonne itself, senior scientists did too little to prepare their young subordinates for big future breakthroughs. Thackeray and Amine— they ought to regularly assign risky and challenging projects to junior scientists "just to try."

"It's like when the lions race the cubs," Kang said. "They push the cubs off of the cliff and see if they survive or not." That did not happen enough at Argonne. And it was shortsighted. Eventually the junior scientists would succeed Thackeray and Amine.

But now it was Kang's responsibility at Samsung.

18

IPO!

Before carrying out fresh experiments, Argonne researchers wrote reports justifying their safety, often twenty pages in length. They wore specified lab coats—in some cases white, in others blue—and large plastic protective glasses. Before putting their hands into glove boxes, they first donned surgical gloves as an added protective layer. If they wished to continue working past seven P.M., they had to obtain special permission. These rules and others were passed down from Washington. Chamberlain pushed for compliance, arguing the tenet that all should go home each day precisely as they arrived. No one complained, at least not openly.

Protective glasses, lab coats, and extra gloves were not mandated at Envia, where researchers experimented when the impulse struck, whatever the time, absent the requirement of any pre-experiment safety reports. Many people thought that its scrappy style was helping to carry Envia into a leading position in next-generation batteries. But Kumar argued that that might be true but that if he was truly to contribute to beating the Asian giants, it was insufficient—he also needed more resources, specifically from the Department of Energy, which he urged to get behind Envia in a bigger way. For the automotive industry to treat you seriously, you had to produce material at scale—in batches as large as a ton. Kumar requested a $30 million loan to build an industrial-scale pilot plant that would produce such quantities and satisfy the automakers.

Kapadia, Kumar's financial backer, said that if, as Obama aspired, there were to be one million electric cars on American roads by 2015, it did not matter *where* they were assembled. Envia would sell to carmakers from around the world, and Obama should support the ambition. The point was that "the cars should simply contain the highest number of U.S. components." That would spur the field. "It would build up the manufacturing base," Kapadia said. The cathodes and anodes of the 1970s, the 1980s, and the 1990s were invented in the United States, Kapadia repeated, but ended up mass-produced elsewhere. Here was an opportunity for the United States to invent and collect the manufacturing rewards on its shores, in a way attuned to the present decade.

Kapadia's notions misunderstood the race. To produce the most advanced, slickest automobile on the planet could arouse passions, but Americans were unlikely to rally around the goal of supplying auto parts. Yet he bluntly advanced his point—if the Department of Energy would not bankroll the pilot plant, someone else would. A player from Japan, South Korea, China, or even Brussels would buy Envia and establish its own plants. "You can set up in China for one tenth of the cost," he said.

Kapadia was advancing a threat—that Envia would make battery components one way or another. There was the risk that, as with the history of cathodes, its intellectual property would be lost to a foreign player.

Halfway through 2011, several sources told Chamberlain that Asian companies had already initiated an informal bidding contest for Envia for just this purpose. They were throwing out large numbers to buy the start-up outright—the reports were in the tens and hundreds of millions of dollars. He was not surprised to hear that Envia was receptive. Kapadia's warning was Silicon Valley reality. Envia was a start-up company and the venture capitalists who had funded it would be keen for opportunities to cash out. VCs typically sought to "exit" an investment no more than five years after injecting cash. Whether or not the start-up had created an actual product tended to be a secondary priority to collecting the fruits of their risk. Some used the unflattering terms "pump and dump" or "hype and release" to describe this aspect of venture capitalism. There was the possibility that Kapadia was exaggerating or out-right bluffing—Chamberlain said he always assumed his interloc-utors across the table were overstating at least somewhat. But he had heard enough directly from the companies to treat the talk seriously.

Around this time, Kumar flashed a slide on the conference room screen. It was from an Envia presentation that had recently persuaded Nissan to sign a $700,000 contract for the development of a customized NMC 2.0 cathode. One slide showed Envia's first-generation version. A typical way to express the economics of a battery was the cost to produce a steady 1,000 watts of electricity for an hour (the amount needed to iron your clothes, for instance). According to Kumar, the Envia cathode lessened the battery cost to $250 per kilowatt-hour at laboratory scale, less than half the prevailing market rate at the time it was built. Envia's next product promised to shrink the cost further—to $200 per kilowatt-hour, a very large jump. The ultimate aim, if Kumar succeeded with a su-perbattery on which he was currently working, would be a phe-nomenal $180 per kilowatt-hour. Kumar told Nissan that he could reach that goal in eighteen or so months. His promises, not to mention the time line, were exceedingly bold seeing as how GM was thought to be currently spending $650 to $750 per kilowatt-hour on the battery in the Volt, for a total of $12,000 to $14,000. Dave Howell, head of the electric-car battery research effort at the Department of Energy, was challenging researchers to lower costs to $300 a kilowatt-hour by 2014 or 2015. His longer objective was $125 a kilowatt-hour by 2022. But Kumar was suggesting he needed a mere year and a half to cut battery costs by three quar-ters and bring down the Volt battery to around $3,000.

Given those numbers, you could understand the frenzy building around Envia—in Asia. American companies were generally am-bivalent about Envia. Kumar found that sad and frustrating. The Obama administration had allotted about $2 billion to build six lithium-ion battery factories largely from scratch. No one could say how many would survive, but most had no intellectual property of their own. In Kumar's view they ought to be eager to grab Envia's battery material. But, hearing silence, he said, "I don't think it's my job to convince them. I am working to make a product."

There was some U.S. interest: Kumar had received a firm bid from one American company—a $125 million buyout offer from A123, the Massachusetts battery start-up—alongside a variety of proposals advanced by German and Japanese companies. But then Envia's management went decidedly cool in buyout discussions. A courtship was heating up with potentially the biggest American battery customer of all.

19

The Car Man

Jon Lauckner was a third-generation Detroit native. His father worked at Chevrolet for three decades. His father's father worked for Fisher, GM's internal body-making division. And Lauckner himself had left Detroit to go to business school at Stanford before returning to General Motors. Slightly cocky, with short, clipped brown hair, fashionable glasses, and a big, carpeted office, he ran GM Ventures, the carmaker's investment arm. In 2006, he took out a fountain pen and notepad and, on the fly, jotted out the design and economics for the 16-kilowatt-hour power train that five years later propelled the Chevy Volt concept car. The plug-in hybrid combined Lauckner's battery with a gasoline-fueled engine, the former feeding a 149-horsepower electric motor. When the battery ran down, the 84-horsepower internal combustion engine took over. Intended to brand GM as a carmaker with the chops to compete on the leading edge of technology and style, the Volt was named 2011 car of the year by *Motor Trend*, which called it "a moonshot." Though it boosted GM's image, the Volt did not actually sell well. The car cost $41,000 and most motorists were unimpressed by the forty miles it could travel on a charge. Studies showed that that was the maximum average distance that American motorists traveled in a day. But in practice, actual potential buyers wanted to pay less, drive farther, and charge up where and when they wanted. Until these benchmarks were met, most were not buying the Volt or any other electric.

The problem was the battery—GM would not disclose the specific composition beyond saying that it contained the first-generation NMC chemistry blended with Thackeray's breakthrough spinel. It was reliable and stable, its cells built by LG Chemical and the pack by GM, and large—it used just ten of its sixteen kilowatt-hours of size. The battery needed that technically idle capacity to conform to how motorists actually drove, including a habit of drawing down the charge until it was almost spent but still insisting on fast acceleration. When a lithium-ion battery was nearing exhaustion, the electrons began to resist extraction, meaning cars became sluggish; a solution was auxiliary energy—the Volt's extra six kilowatt-hours—with which the electrons still moved fast when you pressed the accelerator. Only, the cost of adding the cushion meant that, even if an owner rarely bought gasoline, the savings in fuel did not compensate for the vehicle's elevated price.

Lauckner's calculus was that later generations of the vehicle would validate the economics. By 2020 or so it would be competitive with purely gasoline-driven vehicles. But getting to that year required nerve. "That is the race," said one of his GM colleagues. Lauckner's current aim was to make as much of a technological jump as possible in the next-generation Volt, to be launched in 2016 or so. He needed a much better battery, one that would cost thousands of dollars less and take the car much farther. He believed that NMC 2.0 was his best chance.

That was why Lauckner did something unprecedented. Detroit orthodoxy was to work with big, safe, established suppliers whose products were proven and that themselves would not collapse. But Lauckner embraced Envia, an unknown start-up company, to produce the most crucial component in the next-generation Volt.

The relationship went back to January 2009 when, eighteen months after he and Kumar raised their initial funding, Mike Sinkula managed to obtain a meeting with GM. Kumar was "really scared"—he had only rudimentary lab results and, since he had spoken to no potential customers as yet, no sense whether "I am too early, or too late" in divulging even that data. But "it made sense to show it at least once and see what they tell us."

In a basement conference room in a GM office outside Detroit, Kumar talked through two dozen charts tracking the high capacities that his juiced-up adaptation of the NMC cathode was already achieving, along with what he thought was possible through Envia's version of NMC 2.0. He expected tough questions, but the GM men were "surprisingly positive." A little over a week later, an e-mail arrived from Mark Verbrugge, the head of GM's research and development arm: would Kumar mind shipping some samples of his NMC cathode? Later, Kumar was told that Verbrugge had been to Argonne in the intervening days and checked his claims. Kumar sent the cells.

Kumar didn't hear back for six months. But when Verbrugge at last called, he said GM wanted to negotiate a "joint development agreement" with Envia. The idea would be to collaborate for eighteen months on Kumar's cathode and try to meet a few serious milestones. Verbrugge understood that, unlike the incumbent suppliers, Envia did not have much cash in the bank. So GM would pay $1.5 million for the work. It was an enormously prestigious offer representing "lots of value to Envia," Kumar said. At a California battery conference, Kumar found himself surrounded by his newfound GM colleagues. He noticed sidelong glances from other vendors enjoying no such attention.

The early results from the collaboration were not perfect— Envia persistently failed to satisfy a couple of GM's important metrics. But those it did meet "got GM really excited." Lauckner, who directed GM's high-risk investments, was particularly impressed. Speaking with Kumar, he found him to be a "rocket scientist" on par with or smarter than the best battery guys he had met anywhere. Kumar described "one of the most interesting—if not the most interesting—potential new cathode materials" that GM had found anywhere, Lauckner said. Their performance, if borne out when the material was made in large batches, could make the second-generation Volt a winner. Lauckner said he wanted to make a substantial direct investment. Kumar sensed an aim of turning Envia into a "GM-centric company."

From Envia's side, Lauckner's interest was opportune. The company was essentially insolvent. Kumar had burned through almost the entire $3.2 million with which he started. Bay Partners and Redpoint Ventures had invested another $7 million in Envia after that, but it was almost gone, too. Kapadia was on vacation in India when he received an urgent call from Kumar, who said the firm was nearly broke. "When you wrote the first check, you said, 'I'll stand by you,'" Kumar said. "You need to come and help me." He was holding serious discussions with numerous companies, but none had signed a licensing deal, and the cash had to last until one or more did. "Spend three months with me," he said.

Kapadia cut short his vacation. When he arrived, he found a gloom hanging over the researchers. The talk was of a fire sale of Kumar's ingenious work. Assembling the team, Kapadia uttered a single phrase that all would remember: "We are not for sale." Morale soared. GM was knocking on their door.

Kapadia understood immediately that GM wasn't "just any investor." Just as Kumar's battery material could salvage the second-generation Volt, a relationship with GM could save Envia.

Lauckner proposed that GM Ventures lead a $17 million round of fresh investment in the start-up. GM itself would throw in the biggest share—$7 million. Kumar and Kapadia agreed. The remaining $10 million came from two Japanese companies, Asahi Kasei and Asahi Glass, both big suppliers of battery materials and thus smart strategic connections for Envia. Now there was the cash—and industry credibility—for Kumar to concentrate on his lab work and not mere survival.

Start-ups are unpolished objects, but so was the GM team. Lauckner was not a trained venture capitalist; neither was any member of his team. They "didn't come across as experts or even as seasoned venture capitalists like the [other Envia investors]," said Anish Patel, a Lauckner subordinate. "It was a big learning process from our side." But Kapadia and Kumar found Lauckner to be an astute businessman and, more important, authentic. They trusted him.

The toughest challenge for Lauckner was resistance within GM itself. The most powerful executives in almost any internal battle over GM's fleet were the engineers. They had among the most demanding jobs in the company, since it was their responsibility to ensure that new models were flawless at launch. Consumers, the media, and Wall Street could be ruthless in the event of a serious defect. That was why the big carmakers relied almost entirely on long-standing parts suppliers, large enterprises that, unlike start-ups, had less incentive to exaggerate the capabilities of a single product and jeopardize the sale of their much larger line of merchandise. In the case of volatile lithium-ion batteries, carmakers contracted exclusively with the big Japanese and South Korean chemical companies. Such contracts could be worth many billions of dollars. To win them, the suppliers endured a grueling qualification process often lasting years. The issues were not only consistent quality but the fundamental health of the suppliers—no carmaker would contract with a company that seemed in peril. So their financial wherewithal was reviewed exhaustively.

No start-up could withstand such scrutiny, the engineers said. The cells for the Volt batteries made of the NMC and Thackeray's spinel were supplied by LG Chemical, which, having won the Obama administration stimulus funds, built a factory in Troy, Michigan. The GM engineers advocated another LG battery for the next-generation model, this time containing a blend of the NMC and Goodenough's stable lithium-iron-phosphate. They flatly opposed any licensing deal with Envia.

Mistrust ran deep between the engineers and the venture capital team. The composition suggested by the engineers would be inferior to pure NMC because lithium-iron-phosphate delivered less energy per unit mass. But in the engineers' math it was described as delivering the same performance. That made Lauckner's team think that "some people were manipulating the data to their benefit." From the engineers' side, there seemed to be suspicion that Lauckner was not impartial, either, because he had orchestrated the $7 million Envia investment. A senior LG research scientist encouraged such doubts when he told GM in Kapadia's presence that he would have advised his own company to buy Envia—except that its technology "did not work as advertised."[1] Kumar later counseled Kapadia to ignore such remarks—the South Koreans were merely envious that a start-up might win some of their business. Still, for now, the GM engineers possessed the advantage in the discussions—there would be no license with the start-up.

Lauckner held himself aloof from foreign bidders on Envia. He said little more than that, if Envia *were* to be acquired, he preferred that the buyer be American.

His attitude went deeper than the problem with the engineers. Just two years earlier, GM had itself filed for bankruptcy and required a $49 billion federal bailout in order to survive and regroup. The government came to own 61 percent of the company. Even after a $20 billion IPO in 2010, the government still today held a 32 percent stake. That grated on GM executives and made them ultracautious about new long-term obligations. GM would not buy a battery company, even if it was a clever move, because it was safer to stay lean and focus on building cars and trucks. Lauckner sat on Envia's board and knew Kumar, Sinkula, and Kapadia as well as a business partner could. But there was no bypassing this ingrained opposition to the acquisition of a supply company.

When you asked Lauckner about a potential buyout, he would change the subject. Kumar reckoned that he was gaming events. The GM man was alert: in the case of any formal bid from a foreign firm, he would move rapidly with the objective of keeping Envia in the United States or at least ensuring GM guaranteed access to its batteries. But ultimately, Lauckner seemed to rely on Envia's loyalty should a choice arise between masters. He hoped his early support would pay off for GM.

Kapadia said it was hard to fathom the lack of private American interest in keeping Envia at home. He rang Chamberlain. Was he willing to approach to his American industry contacts and suggest a bid for Envia? Kapadia reckoned that, as fathers of the NMC and the vanguard in the battery race, the Argonne guys would feel duty-bound to help keep Envia—a commercializer of their material—in the United States.

He was right. Chamberlain "felt lucky" to be called on by Kapadia. At his core, he said, "I'm a true believer. I think it would help this country to have a manufacturing footprint in this technology." Chamberlain was also eager for the chance "to prove the case that the stuff we do in the lab can have importance outside the lab." It could be a straight line—from Argonne to Envia and on to a larger manufacturer. If the NMC made it the whole way, there would be personal dividends—"opportunities for me." Plus it would be fun. He said, "I get juice from the deal. It's the rush of playing poker."

With GM hanging back, Chamberlain knew whom to call first—BASF. It seemed a stretch, but Chamberlain said that even though it was headquartered in Germany, BASF, when it was operating on American shores, seemed equal to any domestic company. In fact, three years before, he had already begun pushing BASF and Envia toward this very acquisition. At the time, Chamberlain reckoned that the deal could be done for $20 million. But now the price would be ten times that figure, the number that Chamberlain heard had been offered for the company.

Chamberlain explained to BASF, in addition to two American companies and three potential foreign buyers with whom he had the same conversation, that Envia had executed an exceedingly clever strategy. Rather than marketing a battery through middlemen such as big chemical companies, Envia had gone directly to the end user, which was automakers. Envia lacked the capability to manufacture in volume. Its technology was not yet commercially ready. But it had an open door to Toyota, Honda, Nissan, and GM.

In the industry that Chamberlain knew best—semiconductors—a vendor's most prized possession was a "spec sheet," a precise description of the metrics needed to comply with a company's product

standard. With the spec sheet, you could custom-make a part specifically for a big buyer. But spec sheets were closely held by chipmakers, who feared tricks by competitors seeking to obtain insider
knowledge.

Chamberlain sensed that the carmakers were even more secretive. They were even less likely to divulge their specs. Yet Kumar had managed to obtain the spec sheets of four major global
carmakers. It had taken the Envia founder several years, but he
had ingratiated himself to a point where he intimately understood
the technological needs of Toyota, Honda, GM, and Nissan. All of
them were speaking of commercializing his work—the technology
of a thirty-man start-up. That gave Envia a competitive edge over
BASF, Dow, and DuPont, all of them "machines when it came to
sales, marketing, and understanding a customer's needs," Chamberlain said.

Based on what he was hearing, Chamberlain calculated what
Envia management stood to earn in a buyout. The VCs now owned
46 percent of the company. Kumar's share had been whittled down
to 10 percent; Sinkula's was even less. Still, if the company was sold
for hundreds of millions of dollars, they would be wealthy. Kapadia, too, had been awarded stock as CEO. But when Chamberlain
plumbed the subject with Kumar and Kapadia, they expressed surprise. Why, we've never actually done the math, they said. "We only
want the best success for the technology," they would say.

That was what they *should* tell Chamberlain or anyone else on
the outside, the Argonne man thought. But it was also rubbish.
Kumar and his partners had in fact spent much time on their cal-
culators. Kapadia had hired Goldman Sachs and Morgan Stanley
for independent advice. Both he and Kumar said they favored
cashing out—whenever it was in their interest. But to turn one of
the interested car giants into an actual buyer, they needed to advance their technology a bit more.

They could also take a greater risk and go to the equity market.
They could offer shares in an initial public offering—an IPO, the
traditional aim of Silicon Valley start-ups seeking to monetize their

years of toil. They had a firm idea of what an IPO should raise. A billion dollars, Kumar said.

That was Kumar's and Kapadia's personal goal—a $1 *billion payout.*

Late one night, Chamberlain called Kumar's founding partner, Sinkula. "Look, I am about to put my reputation on the line by pulling the string on this," he said, meaning to stir up American interest in Envia. "You need to tell me whether you really have something that is worth that much money."

Sinkula replied, "Why would Toyota, Honda, and GM be coming to us and saying how good Envia's technology is if it wasn't? I would not jerk you around on this. We have an actual technology."

Chamberlain later said, "At some level, I have to trust that."

He went back to his central motivation, which was "to do something for the United States of America." He believed it would be decidedly disadvantageous for the United States if Envia migrated to Asia.

20

Bell Men

S ome said the problem with batteries was that until recently no one had declared them an urgent need. Neither consumers nor carmakers had said they *wanted* a vastly improved battery. Given demand, the market would have supplied one.

If this was true, all that was required to win the battery race was the declaration of a commercial crisis. Time would take care of the rest. But Peter Littlewood, Argonne's deputy director, said it wasn't. The problem went much deeper than words and economic theology. It was big and required a big answer.

Littlewood primarily blamed industry. Companies that could encourage and develop transformational inventions resisted them because "people go out of business." Sony worked to improve its batteries, but only by a few percentage points a year. The readiness of consumers to accept the current pace of improvement was part of the problem, too. But given industry's tens of billions of dollars in investment and anticipated profit, there simply was no rationale for a big leap. "It is the Microsoft model," he said. "Why write decent software when you can keep selling people upgrades?" Because of such professional self-sabotage, batteries were a disaster. "You open the damn things up, and they are a mess," Littlewood said. "I mean they are an affront to the eye."

Littlewood formerly ran the theoretical physics department at Bell Laboratories. He was part of a cabal within the scientific and managerial leadership of the national labs and the Department of

Energy—former senior Bell scientists and managers who saw their private-sector prior employer as a model for how research should be run. Eric Isaacs, Argonne's director and Littlewood's boss, was also a Bell veteran. The directors of two other national labs were as well, as was the Department of Energy's chief scientist, Bill Brinkman, along with Steven Chu, the secretary of energy, who won a Nobel Prize for work he'd done at Bell.

Littlewood went on from Bell to run Cambridge's Cavendish Laboratory, where in the 1950s Francis Crick and James Watson discerned the structure of DNA. But when he arrived at Argonne in 2011, it was Bell Labs he invoked. If AT&T had been in energy storage rather than the phone business, he said, you might have very different battery technology today. Bell would have spent years if necessary exploring the tiniest fundamentals of battery chemistries and how they interacted. It would have derived a road map—a precise electrochemical latticework of the battery. It would have then methodically ticked off the possible routes to an answer until the puzzle was solved. The proof was in what happened when Bell dabbled in batteries—in the late 1970s and early 1980s, a Bell researcher named Samar Basu developed the first graphite anode, which, improved by Moroccan researcher Rachid Yazami and combined with Goodenough's lithium-cobalt-oxide cathode, became the basis for today's standard lithium-ion battery. "We haven't done that," Littlewood said, "which is why we're in the mess we're in now and why you see scrambled proposals to try to get something that works in a period of a few years." He said, "We're driven because the energy problem is so close to us that we need to solve it. But we don't understand properly how batteries work."

Littlewood turned up at Bell in 1980, just out of graduate school. His specialty was exotic phenomena such as the Higgs boson, the theoretical particle crucial to quantum physics. Bell had no distractions—no teaching, no pesky students—and researchers enjoyed seemingly limitless resources. New researchers would show up and be told, "Do something interesting." Not prolific invention, but do *one important thing* every year. "Now what's important?" Littlewood wondered. The answer he got was, "You can tell when you see it."

You were being appraised against Bell's legacy, which meant the question, "Have you won a Nobel Prize? Have you invented some new method of doing something?" That was the standard through Bell's three quarters of a century of history. "There were ten thousand scientists at the heyday," Littlewood said. "They were not all Nobel laureates, but were all at a very high level. A bunch of arrogant bastards, all of them." But as self-assured as they were, Bell scientists also realized that, because they were aiming at the truly big breakthroughs, they needed the help of colleagues. They cooperated.

If you looked at Bell that way, you realized that the lab had been assembled rather carefully, with different kinds of talent—all individuals, all excellent at their specialties—put in an environment where they could not afford not to interact. Because AT&T was a regulated monopoly, it exerted little pressure on its Bell unit for a commercial payoff. Company managers hoped that any particular piece of Bell's research would prove useful perhaps decades down the line. So while the pressure was intense to produce first-rate science, there was almost no insistence that expenses be justified from a business aspect.

The atmosphere discomfited some researchers. "Every year they'd hire a bunch of kids who would work eighteen hours a day," Littlewood told a group of visiting battery guys, and the veterans would have to compete with them, too.

"Did ideas get stolen?" asked Venkat Srinivasan of Lawrence Berkeley National Laboratory.

"Of course."

As for Steven Chu, he felt like a member of the "chosen ones" when he joined Bell in 1978. The atmosphere was "electric," and "the joy and excitement of doing science permeated the halls," he said. Chu grew up on Long Island, the son of Chinese immigrants who expected their children to earn Ph.D.s. His maternal grandfather was an American-trained engineer. His father was an MIT-educated chemical engineer and his mother an economist. He earned his doctorate at Berkeley and was hired to stay on as an assistant professor, but before starting the job he was offered a leave of absence to broaden his experience and he used the time to go to work at Bell.

Chu's first Bell boss admonished him to be satisfied with nothing less than starting a new scientific field. Five years later, he was leading the lab's quantum electronics research team. Among his first accomplishments was measuring the energy levels of positronium, an atomlike object with its electric charges flipped. Measurements were hard because positronium has an average lifetime of 125 picoseconds (125 trillionths of a second, a scale that is to a second as a second is to 31,700 years). Then Chu puzzled out how to use laser light to cool and trap atoms. "Life at Bell Labs, like Mary Poppins, was practically perfect in every way," he said.

As secretary of energy under Obama, Chu wanted to capture the magic of Bell and its peers, the great industrial labs that had been run by scientific and commercial visionaries like Thomas Edison and T. J. Watson. He wanted to assemble the best minds in one place and focus on a single mission. The objective would be to disrupt the largest industry on the planet—fossil fuels.

A half century back, such an approach to business was part of the American DNA. But today, Intel, for instance, while still on top after decades in semiconductors, had narrower aims. "They're not going to do the stuff like Bell did," Chu said. Neither were universities.

Chu wanted to establish long-term research programs that, while undirected as to their specific product, would almost certainly emerge with important and most likely foundational results. At Bell, Chu said, you learned to take on "a problem, think about it hard, solve it, write up the paper, submit it, and move on to the next one." He wanted a similarly fast environment.

Bell sought to elevate the best scientists to management. Given authority, they were held responsible for the productivity of those under them. There was no peer review. Some scientists called the system "su-PEER-ior review," Chu said. The culture kept the checks and balances. If you made a bad decision, a community of great scientists was on hand to step in to help fix it, "but not because they wanted to be the boss." It was, "You've got an idea? Come. Let's go to the board. Let's talk about it now. 'Bing, bing, bing, bing, bing,'" Chu said. He himself could be an exacting boss. When he later was named director of Lawrence Berkeley National Laboratory, he became known for his "Chu-namis," stormy fits of pique when something had not been carried out to his standard. Chu wanted to replicate this atmosphere at the national labs that the Department of Energy funded.

Littlewood cautioned Chu that there were bits of Bell that you would not wish to copy. AT&T was very good at its reliable phone business. It had some of the world's best fundamental scientists, who thought ambitiously about the future. Bell employees were proud of their technological leaps. But the company often did not drive its breakthroughs through to an actual product. AT&T earned only nominal licensing fees for the transistor, for example, though it was invented at Bell. Its scientists conceived the first cellular phones in 1947 and, a quarter century later, the system of transmission towers through which they work, but others pioneered the mobile phone business. AT&T had a landline monopoly and gave away its other inventions as a price of peace with regulators. It worked as a business strategy, until Ronald Reagan's Justice Department aggressively pursued AT&T's breakup. That left it in pieces, absent the patents and side businesses that could have fueled its survival.

Littlewood said Bell was commercially flat-footed. Unless efforts were fully deliberated, he continued, long-term achievements would be limited. He saw Bell as an example of unfulfilled potential, like the Apollo Moon mission. Had Apollo been better planned, he said, "we'd still be there." "It turned out to be an interesting technology program that, if you were thinking ahead and not just to 1970, you would have studied the fundamentals and taken it to Mars." You would not aim for a victory that meant simply stepping your toe on the Moon a few times. In numerous cases, the declared goals of starry-eyed American politicians went unmet. The Apollo narrative promised big leaps if only the goal and budget were pledged. But it was a singular event. Apollo could not be replicated by force of will.

The national labs had originated as entrepreneurial places in World War II. Eventually numbering seventeen, the labs had lost their spunk in the intervening five and six decades. At Argonne, a scientist often would mention a certain chemistry that seemed promising, but when you plumbed further you would discover that he was not actually working on it. "That's not funded," he would say. It had not been sanctioned and paid for by the Department of Energy. When a battery guy would invoke that excuse, you would presume he or she was joking because it sounded so pathetic. But you would see it was not a joke. The truth was that the labs lacked the system, and the scientists the mind-set, for the rapid pursuit of hot new ideas.

The federal research effort had devolved into an assemblage of disparate projects, each worth $300,000 or so. This gave the labs the feel of aimless institutions. No one vetted battery projects from the top, ensuring that the outcome was at least one very good battery *system* rather than a dozen wonderful electrodes and an unrelated electrolyte. The philosophy seemed to be that somehow the better battery would all come together of its own accord. But it hadn't—not yet anyway.

Chu's idea was to overturn this system. He would cut out the "principal investigator," the main scientist answerable to almost no one before all the money was spent. In his or her place would be team-funded work grouped in blocks to attack big problems identified either from below or from above. If scientists veered off on their own, they could be halted in their tracks because funding would not follow them. In theory you could corral people into line and achieve more coherent results.

You could recreate the major industrial lab.

Chu considered China's method. What China did, he said, was to identify a known technology and, in a twist on Japan's approach, squeeze efficiencies out of it by methodically, incrementally improving it over a long period of time. By doing that continually, say, for a decade, China would end up with a dramatically different technology. Not something of the scale of the integrated circuit but, stretched out over a few decades, a very, very good result—"unbelievable," Chu said. "Revolutionary."

If you looked at the leading ideas in batteries, America's were of a different order from those in China, Chu said—they were embedded in products, while China's innovations were by and large accumulations of tweaks on others' work. The lesson of history was that you could not be satisfied with that lead. Quite apart from the battery debacle, the United States invented the airplane but soon lost the lead to French and other European inventors. Germany invented automobile manufacturing but was overtaken by Henry Ford.

Chu intended to follow the latter example. "We're gonna lead in batteries," he said.

21

The No-Start-up Mystery

Chamberlain could not forget a conversation he had back in 2006. He and a lab manager were admiring a new piece of prototyping equipment that could create the large batches of cathode material demanded by industrial customers before they would consider licensing a patent. Chamberlain, then new to Argonne, imagined that the equipment would prompt greater risk-taking innovation by the researchers, now able to demonstrate their inventions in a format that companies understood. It would play to Argonne's advantage. "I imagine that your scientists take risks all the time," Chamberlain said.

Perplexed, the supervisor glanced at Chamberlain's direct boss. Then he said, "Oddly, it works the opposite of what you'd think. Yes, the jobs are safer than in industry, but the job is so good, why take a risk at all?"

Over the next couple of years, Chamberlain witnessed the problem himself—Argonne researchers suffered from a mortal fear of screwing up, or simply looking stupid, that often trumped the desire to make a big splash.

The aversion to risk did not seem to flow from the top. Three consecutive presidents—Bill Clinton, George W. Bush, and now Barack Obama—generously funded lithium-ion research, spurred on by optimism about electric cars. But Chamberlain said that, along the way, a couple of Department of Energy program managers were "whacked" for their risk taking. Their careers had been derailed. Memories of those bruises were still alive at Argonne and might have partly underpinned the hesitancy to gamble.

Given Amine's inclinations, it was easy to imagine *him* wrestling with the temptation to make the jump and form his own company. But he discouraged this line of thought. "If I take one of the technologies that Argonne invented and spin it off and make a company, I will be successful. I'm pretty sure," he said. "But if you move completely to business, you are more likely to be product-focused—you will focus on money. Your innovative brain will go down. It's not my style. . . . I'm not really motivated just to earn money." He said, "I'm not extremely wealthy, but I'm doing fine. A middle-class guy." Amine was entrepreneurial within the confines of the lab. He was proactive and aggressive. But he lacked an appetite for the outside gamble.

Thackeray's impulses were similar. He said he never contemplated a leap such as Kumar's—raising venture capital, licensing his own technology, and building a business around it.

It was the same with the rest of the battery guys—and really the entire lab. Unlike their entrepreneurially frenzied university contemporaries in Austin, Boston, and the San Francisco Bay, the Argonne guys had no record of turning their ideas into profit-making enterprises.

But the option existed. In 1980, two American senators—Birch Bayh and Robert Dole—pushed through legislation that gave federally funded universities and laboratories the right to profit from their research. The Bayh-Dole Act was a response to the listless economy of the 1970s. American inventiveness meant the economy could eventually persevere, but the thinking was that it would help if scientists were motivated by personal profit as well as national honor. Prior to Bayh-Dole, federally funded researchers received no profit share from their inventions. Now they could. The precise cut depended on the institution. At Argonne, the allotment was 25 percent of licensing fees and royalties to the scientist or scientists and, as recognition of the general effort, the remaining 75 percent to the department and division in which they worked. That was how Thackeray, Amine, and the handful of other battery guys had pocketed more than a million dollars from the NMC patents. Yet by and large, the battery guys were timid.

It turned out that not just Argonne but the nearby university as well seemed to ignore the value of its intellectual property. Over the years, researchers at the University of Chicago had won fifty-five Nobel Prizes. Just one discovery—erythropoietin, a hormone used to treat anemia in dialysis patients—would eventually earn Amgen some $40 billion. But neither the university nor its discoverer, a researcher named Eugene Goldwasser, would earn any royalties, as the university hadn't patented it.

In 1986, two graduate students from the University of Chicago moved into offices at Argonne. Clint Bybee and Keith Crandell were volunteers in a new initiative to shake up the sleepy lab and university and cultivate some start-up businesses. They called their firm ARCH, targeting unseen inventiveness at Argonne and the University of Chicago. Starting from scratch, the two men— both twenty-six-year-old business school students—would wander the long halls and collar a scientist. "Tell us what you're working on," Bybee would say. But after a number of fruitless encounters, Bybee came to understand that the Argonne guys simply "didn't think of commercial applications for what they were doing very much." They by and large aimed at the tastes of their principal funders—the Department of Energy or the Pentagon. "That's who their customer was," Bybee said. The ARCH men unearthed a dozen ventures, including four at Argonne, over the subsequent few years. They came up almost empty-handed.

As for the battery guys, they knew the ARCH team was around—they gathered it was a high-level scheme cooked up by the lab director and university officials. But almost nobody came face-to-face with either Bybee or Crandell. Bybee said neither he nor anyone else at ARCH found any battery technology sufficiently interesting to pursue, then or ever, and never took a glance at the NMC. "I never met them," Thackeray said.

The battery guys were feeble entrepreneurs compared with their

Silicon Valley counterparts, and so was Chicago's VC community, even when embedded right in the lab. Sujeet Kumar and companies from Germany, Japan, and South Korea grabbed the NMC while the ARCH men were present but looking the other way.

Chamberlain had a stock of stories from his pre-Argonne days. He knew enough science to hold his own with the scientists but seemed to deliberately steer clear of their projects. He was there to create the conditions in which they could produce their magic and then marshal it into the market. Apart from the occasional pointed question that demonstrated he knew his stuff, he left them to the battery work. He was not really a "battery guy" was how he put it. Thackeray would repeat his flattering assertion that, no, by now Chamberlain *was* a battery guy; he had been around long enough. But that was just Thackeray as a gentleman. Chamberlain wrung his hands over his paucity of actual time in the laboratory. His bench time had been short—at Georgia Tech and a bit here and there in private industry jobs. Could he himself invent something new or write a profound scientific paper that drove peers to sit up? No one could say, and he had never put himself on the line to find out. Yet Chamberlain occupied a supremely respected place in the lab, anchored in his perceived appetite for risk and his grasp of business.

He was different from those around him.

Chamberlain regarded his entrepreneurial inclinations as "something genetic and probably environmental." All his brothers had at one time or another started a business—one in fiber optics, another in chemistry, and the third working with their father at JA Chamberlain and Sons, his Florida marine engine shop. A string of companies had exploited Chamberlain's people and leadership skills and his start-ups had gone some distance. When he arrived at Argonne at age thirty-nine, he soared. There simply was no one at the lab—apart from Khal Amine, and he was regarded with great suspicion—who was equipped with the salesman's bone. So that when Chamberlain did what came naturally—ringing up

senior managers of the major companies and winning them over—
he became the most successful patent representative not only in
the unit, but perhaps in lab history. For Argonne, he was the guy
"from industry."

He was largely tapping his formative experience in semiconductors. Although legend drew a straight line from Bell Labs to Silicon Valley to the iPhone, by the middle 1980s the American
semiconductor industry was in fact dying. It was part of the narrative of industrial decline and the Japanese juggernaut that so con-
sumed Americans. To fight back, American chipmakers proposed
an experiment. They would band together along with the federal
government and attempt to leapfrog the Japanese. In 1987, Ronald
Reagan signed legislation that embraced the experiment. The law
created Sematech, for Semiconductor Manufacturing Technology.
Fourteen chipmakers and DARPA, a Pentagon research arm, went
fifty-fifty on a five-year, $500 million effort to keep semiconductor
manufacturing in the United States. American chip making surged
back. Intel, led by Andrew Grove, regained dominance with first-
rate, intricately designed microprocessors that captured the lucrative high end of the market.

After that triumph, Sematech became a paradigm like Apollo
and the Manhattan Project, shorthand for how industry and government could collaborate to recapture a market from a foreign
upstart.

In 2008, Chamberlain proposed the idea of a battery Sematech
to a Department of Energy supervisor. The consortium he had in
mind would not *save* American battery making per se since virtually no lithium-ion capability had ever been established in the
United States. Instead it would unite companies that together
would *create* an American lithium-ion industry—battery start-ups,
chemical makers, lead-acid battery makers, and so on. It would
put the United States on a footing to compete with global battery
makers. The supervisor understood Chamberlain's idea and challenged him to "get industry to do it."

And so he did. In spring, he invited the executives of a half

dozen battery and chemical companies to a couple of meetings. At
first the companies did not bite—notwithstanding the chipmakers'
positive experience, they wondered how they could possibly collaborate when they were rivals. None liked the idea of a five-year
commitment, as Sematech had been. Yet they gradually coalesced
around Chamberlain's main idea and morphed it into their own,
with the aim of establishing a foundry where battery teams could
prove their materials, a single plant where they would all make
lithium-ion batteries.

One day, Jim Greenberger, an outside member of the group with
which Chamberlain was speaking, mentioned a vague boyhood link
to a close ally of Senator Obama, whose presidential campaign was
gaining momentum. Obama seemed to be intensely interested in
batteries. Why not pitch the battery Sematech proposal to the senator's team? Everyone agreed that it was a good idea.

The group found itself in a Chicago office before a single economic adviser to Obama. Greenberger described Sematech and
the aim of beating the big Asian battery makers.

"Why do you think we can compete with the Japanese auto
industry?" the adviser asked.

Chamberlain said American companies, while currently strug-
gling, could recover and figure large in a reconstituted global
industry. But he added that if electrics truly took off, Detroit, with
its record of stodginess, "will go the way of the dinosaur." They
would not manage the transition to the new world.

"What kind of money do you need?"

The group had discussed this question. If they were modeling
on Sematech, the sum should be around $500 million. But they
wanted a cushion in case expenses were higher. So they decided
on $1 billion. It was perhaps a hubristic price, but that was what
they would request for the battery Sematech.

"Two billion dollars," Greenberger said.

The rest of the group went quiet. Chamberlain could not see
the expression on the Obama adviser's face, and no one could
fathom the origin of the new number.

"Okay," the adviser said.

Outside, the group laughed. Why did Greenberger double the figure? "I don't know," he said. "It just felt right."

As Obama was elected, the economic landscape transformed. The world was in financial collapse and the country in a panic. On taking office two months later, Obama quickly proposed, and Congress approved, a $787 billion economic stimulus package. It was meant to rescue the economy and plant the seeds of future industries. Chamberlain smiled as he studied the breakdown of spending. It included a $2.4 billion line item—a $2 billion lithium-ion battery manufacturing program plus $400 million for the development of electric-car–manufacturing processes.

Rahm Emanuel, Obama's new chief of staff, had remarked that, politically speaking, no crisis should go to waste. The battery Sematech was a "go."

It was and it wasn't. The money would fund the creation of an American lithium-ion battery industry, just as Chamberlain and the companies envisioned. Only now, with the unexpected largesse of a $2.4 billion research-and-development fund, the companies changed their minds about working collaboratively. Johnson Controls received $249 million of the fund, EnerDel won $118 million, and $200 million went to A123. They would compete against one another for the market. There would be no battery Sematech—no industry-government consortium. But the United States would be in the battery game.

Steven Chu also saw no reason to squander the crisis. In his case, there was the matter of his dream to recreate Bell Labs. He proposed eight projects, each tasked to solve a single big problem, at a total five-year cost of $1 billion. For those who did not grasp the significance, he said, "We are taking a page from America's great industrial laboratories in their heyday." On paper, they would be called "innovation hubs." But more explicitly, they were "Bell Lablets."

One of Chu's hubs was to be aimed at revolutionizing batteries. As impressive as NMC 2.0 was compared with its predecessors, it couldn't power an electric car competitively with the internal combustion engine. After accounting for the loss of energy in combustion, a kilogram of gasoline contains 1,600 watt-hours of stored energy. State-of-the-art lithium-ion batteries, by comparison, delivered about 140. Thackeray's goal for NMC 2.0 was to double current performance plus cut the cost. But even that would leave batteries still about a sixth the energy density of gasoline. The Battery Hub's goal was to make the next big jump after lithium-ion—to 600 or 800 watt-hours a kilogram. Toward that goal, the Battery Hub would receive $25 million of federal funding a year for five years, $125 million in all. A competition would decide which university, national lab, or consortium would host the Hub.

Chu advised that those interested stay tuned as to when the competitions would take place.

John Newman, an electrochemistry professor at UC Berkeley, phoned Thackeray. Newman was an icon who had written the standard university textbook on electrochemical systems.

"Why don't you lead the Battery Hub and we'll do it with you?" Newman said.

The competition had not yet been announced, but Newman was suggesting an interesting head start. He wanted Argonne and Lawrence Berkeley National Laboratory, traditionally bitter rivals in the battery space, to submit a joint bid. The approach was surprising given the jealousy between their two institutions. Argonne and Berkeley *never* worked together. They harbored a deep well of mutual suspicion. The stakes, however, were enormous—whoever landed the hub would be the undisputed center of American battery research. Therefore, if they joined hands, agreed to divide the research funds, and did not quarrel, Berkeley and Argonne might stand an improved chance of winning the competition.

In June 2009, Newman traveled as part of a Berkeley group to Argonne. Crowded into a small conference room, they began to brainstorm what a Battery Hub would look like. So much was already going on in the field—depending on the year, the Department of Energy alone was spending $50 million to $90 million on battery research. What could a hub add? Someone suggested starting over—that they wipe the whiteboard clean and simply construct a chart of a first-rate, industry-leading battery research program. They could then shade in areas where there was already sufficient work. What remained would be the proposed Argonne-Berkeley Battery Hub.

The result was a blockbuster, over-the-top plan for a $100-million-a-year, multiyear partnership of companies and scientific institutions. On paper, it was four times the size of Chu's hubs. Both teams loved it. When Chamberlain described it quietly to a few industry friends, they seemed equally enthusiastic, making clear they were prepared in principle to share the cost fifty-fifty with the Department of Energy. Chamberlain thought he understood the companies' eagerness. It wasn't that it looked like Sematech, although the resemblance to Chamberlain's obsession was more than passing. It was because "it was like Bell," he said. Genuinely like Bell, and not the lablets that Chu was proposing. The Argonne-Berkeley team called it the National Center for Energy Storage Research, which they pronounced "En-Caesar."

Congress had to directly approve such spending, and it treated Chu's proposal with skepticism. Its 2010 budget funded just three of the eight innovation hubs. Worse, it guaranteed the money for only a year rather than five and allocated $22 million for each hub instead of the proposed $25 million.

The Battery Hub did not make the cut.

It was a letdown. From Washington, Chamberlain's supervisors assured him that the Battery Hub would be approved the next time. But there was no foundation for confidence. The politics might continue to flummox Chu.

The Argonne-Berkeley group kept talking. Mostly it continued to lay the groundwork for En-Caesar. But it also strategized for the smaller Hub. In connection with the latter, Chamberlain and

Newman traveled to Washington to pitch their intended collaboration.

Newman told Chamberlain to prepare for a flabbergasted reception. There simply would be disbelief in the Department of Energy. Chamberlain thought Newman was exaggerating, but sure enough, sitting there before David Howell, the head of the Department of Energy's battery research effort, he saw that the Berkeley man was right.

"You guys are here together," Howell said.

"Yeah," both Chamberlain and Newman responded.

"You even got John here," he said to Chamberlain.

"Um-humh."

"So you guys must be serious about collaborating."

Howell seemed unable immediately to move beyond their history. The two men went back to their respective labs. But a month later, Howell rang Chamberlain. One of the slides he and Newman presented had described an industrially popular method for saving costs while driving up inventiveness: you crosshatched your talent pools on a matrix so that, say, a single materials sciences unit could service three separate battery teams. All the big companies used this system, Chamberlain had said, interlacing basic and applied science groups. Howell told Chamberlain that senior Department of Energy management liked the idea so much that, picking one quadrant on the matrix, they wanted them to try it out.

"For a million bucks," Howell said.

The quadrant attacked the problem of ultraenergetic lithium metal: that, when you equipped a battery with pure lithium metal, it could burst into flames on contact with air. It was what killed Exxon's original battery in the 1970s. But extracting lithium metal's innate energy efficiently, safely, and durably was an enduring goal, so Howell wanted Chamberlain and Newman to take it on. If they could overcome its volatility, they could enable the next generation of powerful batteries.

Howell's offer amounted to an explicit dare. If they wanted

their $25 million, they had to first solve one of the hardest problems in battery science.

"You say you want to work together?" Howell asked. "So show us."

They took up the challenge. Chamberlain put one of his best thinkers on the task—Jack Vaughey, a chemist whose other advantage was an unprepossessing manner that allowed him to work well in teams.

Even led by Vaughey, lithium metal bogged down in five months of talks on the legal details of a joint program. Chamberlain and Newman could see they had been outwitted. Who were they kidding: $25 million to sit in traditionally antagonistic labs 1,900 miles apart and, what, *lead the way to a new battery age*? When, even with the impossibly unthreatening Vaughey in charge, they could barely launch a single hard investigation?

But if Howell's test was somehow fruitful—if they *could* smooth through their ego-driven testiness, if they could demonstrate progress with lithium metal, perhaps they would prove him wrong and gain an edge in the expected Hub competition.

A few months later, Chu's science adviser, Bill Brinkman, stopped by Argonne. The Bell veteran put aside a couple of hours to speak with Chamberlain, Thackeray, and Amine.

Brinkman stared as Chamberlain threw slides up on the wall describing En-Caesar. Then he broke in and pointed.

"That's the only way we will be able to catch up to Asia," Brinkman said. "That is the only way we'll do it."

For months, there was continued silence from Washington. Politics had shifted. Congress, already impatient with Chu's big ideas, defeated climate change legislation, which had seemed a shoo-in after Obama's election. Chamberlain's group grew gloomier.

Argonne director Isaacs called Chamberlain around Christmas. Screw 'em, he said. "Let's not wait for Congress and the Hub. Let's pull together En-Caesar."

Isaacs was giving the green light for the massively more expensive and ambitious project. Chamberlain should proceed and formalize industry funding commitments. Argonne and Berkeley would build this Bell-like facility without Washington's help, he said. The way Chamberlain heard the gist was, "We are going to collaborate, and we are going to race with the Asians and we are going to win this race."

Since the 1990s, Japan, China, and South Korea had built themselves up as virtually the sole makers of powerful batteries. Brinkman said that, in order to break their hold, the United States needed the factories awarded funding by the stimulus. The United States had to manufacture the advanced products that would flow from the inventions at Argonne and other labs. One capability fed off the other—when you could invent, you could manufacture, and when you could manufacture, you could invent. They were symbiotic, and to have just one of the two would gradually erode your abilities in the other. But even as the United States built its manufacturing capability, other trends had Brinkman worried. China was developing "a super-cadre of people," he said, training them at Tsinghua and other great universities. Foreign students still tended to make up half the enrollment in American graduate programs, and 60 or 70 percent of them stayed and made their careers in the United States. But as China evolved, wouldn't at least some of its graduate students simply stay home—wouldn't the trend of retention so far at Argonne suddenly go in reverse? If you could attend a first-rate university in the middle of Beijing— if you were the best student in the class and were treated as such—"where would *you* go?" he asked. Brinkman believed that the United States would begin to lose such students to China. They would stay home and present a new challenge in the battery race. That was why En-Caesar was probably the only way to win.

22

"The Damn Hub"

E ric Isaacs had ordered Chamberlain to proceed with En-Caesar, but both men still hoped and expected that Congress would approve a Battery Hub. Presuming they were right, their idea was to first win the Hub and later make it the cornerstone of En-Caesar.

If there was going to be a Battery Hub, Isaacs badly wanted to win. A victory would confirm Argonne's battery preeminence. But the game was high stakes because a loss would be no small matter—it could devastate the Battery Department's methodically constructed reputation, and that of the lab itself.

The latter prospect preoccupied Isaacs and his lieutenants much more than the former. Argonne seemed cursed when it came to large projects, the defining legacies of great national labs and universities. Since 2008, Argonne had entered and lost competitions to host three major federal research facilities, including two of Chu's early innovation hubs. Considering Argonne's origins in the birth of the Atomic Age, the toughest blow was losing a nuclear modeling hub to Oak Ridge National Lab. "When we didn't win, you could have knocked anyone over with a feather because we were so sure," said an Argonne staffer. "There was questioning whether we would even continue to do nuclear research."

The stakes with the Battery Hub were even higher. If Argonne, with its long history in batteries and the presence of Thackeray and Amine, could not win the Battery Hub, it would be an unfathomable blow. No one wanted to speak of the consequences that would follow. The researchers whispered that the senior lab leadership—Isaacs and his deputies—would probably all be fired. Chamberlain would not lose his job but his aura would be darkened and his rise at Argonne stunted. As for the Battery Department, the researchers spoke starkly of an ineradicable stigma. The lab would seem unchanged at first. But slowly it would wither away. Existing grants would go unrenewed. Big, new grants would not come.

In February 2012, an announcement appeared on the Department of Energy Web site: within ninety days, anyone interested in hosting a new innovation hub for energy storage—a Battery Hub—should submit a proposal.

Isaacs had another order for his subordinates: "Win the damn Hub."

Amine—the lab's wheeling-and-dealing, big-thinking risk taker—didn't think Argonne would lose. He thought it would win the Hub and then rapidly pivot to En-Caesar. En-Caesar, which did not entail competition but savvy and nerve, meant much more money and an elevated opportunity to decipher the battery. He calculated that if ten teams of scientists backed by a lot of money worked on the electrochemical and physics challenges in a consolidated manner, the big leap "will go fast."

This combination of grand, practical thinking was the glue between Amine and Chamberlain. Both had worked in industry, Amine in Japan, Chamberlain at American chemical companies. Both knew "that any innovation has to move," as Amine put it. It could not be filed away because it would soon be forgotten. Amine was just then aiming inventions at two companies and was confident both would provide funding to help develop them further. This was Chamberlain's proactive approach with En-Caesar, too—he did not wait but went out and obtained informal industry funding commitments.

That was how you survived. You needed a stream of funding, without which there was barely any use in even conceiving an idea.

Government or industry investment—both had to be fought for. So you collaborated with someone rather than not working at all. You sought connections with funders constantly and you delivered on your promises in order to keep the flow going.

The two big projects—the Hub and En-Caesar—were not only about surviving. In the view of both Chamberlain and Amine, the powerful combination of the two could win the long game against the Asian giants.

Both men were also contemplating the personal payoffs to come. Chamberlain was already superlatively ambitious, but the Hub offered another dimension of opportunity. A battery guy in the Bay Area had remarked that if he managed to oversee the big leap in batteries, he would be made in Silicon Valley—everyone would know his name. That notion—that he might "stand out in Silicon Valley"—fixed Chamberlain's attention. He had failed to do so when he was raising money for his start-ups, but his Bay Area friend might be correct that the Hub could finally attest that he had done it before.

23

Team Argonne

S cientists had been working for two centuries on the battery. It was a hard problem. If you did not believe success possible—if you would not sit down with your colleagues and work it out in the open—a super-battery would never happen. Chamberlain was certain that it was possible. But he had to transform the department into a team.

The push for collaboration went from him, to his Argonne bosses, up the chain to and including Steven Chu, with his dream of recreating Bell Labs, and, when you considered the nation's priorities, all the way to Obama. But for a lot of the battery guys, it was still hard to foresee how the mantra of teamwork would gel as a practical matter. Successful collaboration simply didn't add up to hard proof of inventive work done well.

The idea of a system that glorified teamwork grated on Thackeray. It was not a natural state in the pursuit of science. It was "being thrust down onto us by DOE," an enormous management mistake that did not take account of "the odd people" who populate labs everywhere, those who "probably operate a little bit better in their own little orbit." In Thackeray's view, the team system was going to produce new varieties of the same, preselected battery chemistries while probably failing to mine "all the other stuff that's lying on the outside." To get at potentially breakthrough unconventional batteries, you needed "enough time to develop a new idea and see how it goes."

In Thackeray's experience, scientists tended to be individuals—individual thinkers—and highly competitive. They flourished only in the right environment, which meant recognition when they were successful. "If you give your idea and then somebody else just takes it and runs with it, what are you left with?" he said. Chamberlain had to convince Thackeray and other lab scientists that their intellectual property would be husbanded and not subsumed into the maw of general ideas—or be outright stolen. Otherwise, Thackeray said, you would end up with "people not trusting one another and making outrageous sorts of remarks."

Thackeray meant the accusations of theft that led to Amine's inclusion on the NMC patent. But he was also referring to his own intention to stay outside the team system regardless of what Chamberlain and the Department of Energy desired. When Chamberlain's deputy, Tony Burrell, attempted to promote Chamberlain's teamwork mantra, Thackeray dismissed him. "You know," he said, "he's not a battery guy. He's still got quite a lot to learn."

Perhaps the teamwork idea would have been accepted without question a few years back. But Thackeray thought that the NMC licensing success had irreversibly changed Argonne's culture. "It's unfortunate that money talks," he said. But the fact was that now "everyone would like to become part of this licensing." And that reinforced the notion of individualism.

The Hub was an important aim, and teams and collaboration sounded worthy. Doubts gnawed at some of the core guys underneath Thackeray and Amine as well, men who had been around the lab for two decades, years in which success meant not collaboration but shining in individually executed projects. Would there be a patent or a paper, one with your name in the lead position, the only place it counted? If not, as seemed likely, how would you explain yourself in the all-important "Annual Merit Review" in Washington, the official vetting of every scientist in the system, where, before large audiences, you justified your existence in the laboratory? How would you validate your work? "Ummm . . . I've been collaborating with those other thirty guys." It did not seem a credible proposition. It went to the bedrock of modern Western tradition, which was to reward individuals. The Nobel Prizes, after all, went to individuals or up to three-person teams, not to sprawling groups. You could not change this psychology by fiat.

Chamberlain was never going to change Thackeray and Amine. They had too much leverage and could and would simply say no. Because of their reputations, they would escape any repercussion, too. Although others in the lab were on less solid ground, some of them felt the same. At one meeting, Daniel Abraham, a lab veteran, said flatly that he was not prepared to distribute his data openly to the department at large. What was his advantage in doing so?

Yet that was precisely what Chamberlain was proposing. Everyone needed to think big. The Hub needed to be about teamwork. He was going to "bring this up and hit it right on the head."

He summoned a grand departmental meeting in the basement auditorium.

"I want to start by reminding everyone of the mission we're on," Chamberlain said, pursing the fingertips of his hands. "And that's the electrification of vehicles, and then the extension of that would-be improvement in efficiency to the electric grid." Standing before a full auditorium, he said the mission had three components, all of them anchored in security.

First was economic security. Rechargeable lithium-ion batteries produced more than $10 billion in global revenue. As the vehicle market grew, the market for lithium-ion batteries would exceed $100 billion—by most people's estimates, that would happen within five or six years. These estimates excluded manufactured materials used to make the batteries, not to mention the market for automobiles themselves, which was enormous. Companies elsewhere in the world—in China, in Japan, in South Korea, for example—were gearing up to capture that business, to create jobs for their societies. That was one reason Argonne was on a mission to create a lithium-ion battery industry in the United States.

Another was energy security. "So we import a lot of oil, a little over a billion dollars a day," he told the silent room. "Some of that is from friends, some from enemies. So that's not a secure position." And the third component was environmental security—if the United States could move away from fossil fuel consumption, climate change would be less of a threat as well.

"So in the big picture," Chamberlain said, "I wanted to remind you of the mission we're on." Altogether, he said, this mission was in the service of an investor—"the American taxpayer."

You were not required to join this mission if you preferred not to—it was not an obligation of employment at Argonne. Not at all. But he did not then—nor in any of the other dozens of times that he offered his researchers a way out—sound convincing in this respect. He was calling for an unequivocal commitment. Either you were in or you might be out.

The mandate sounded terrifying. There were historical examples of group glory in science, but they were few, grandiose, and by now trite because of overuse—the Manhattan Project, Apollo. But it was up to Chamberlain to prod the group to trust that the team approach he planned was not a cliché. That they should be calm as they went along. Their careers and their chances for scientific achievement would *not* be derailed. Instead, they would have greater opportunity to do big things and be recognized for it.

Chamberlain seemed to sense that he had the researchers with him. The way he described it—the big science that was going to be done in teams, the big breakthroughs that might be made—why, that was why they *were* in science. He went on.

If there was a Hub competition and Argonne won it, he said, "we could really catapult ourselves" into a position to achieve the mission. Just a week earlier, Argonne had conducted a "Director's Review," in which the heads of other national labs stopped by to review the preliminary Hub proposal. "I don't know how to say this exactly," Chamberlain said, "but they were really excited. They were using words like 'blown away,' and 'ecstatic.' And I will tell you, I was nervous. I wasn't sure. The biggest weaknesses that were identified were the way we were portraying the message—the way we built the story, which is very different from the guts of the science, which they were happy with. We're still, in my view, way ahead of the game."

If Argonne did win, more and more companies would seek a piece of the multibillion-dollar battery market. They would realize they lacked the research capability. They would look around, kick over stones, and then turn to Argonne.

"That's what we are in the middle of right now," he said. That was what collaboration could bring to Argonne, and to the country.

Fire

A h, crap," Chamberlain said. He was reading a news bulletin on the Internet—a Chevy Volt had caught fire while undergoing federal crash testing in Wisconsin. The vehicle had been through the usual harsh examinations, which included ramming a pole into its side, and had already achieved the top five-star rating. Three weeks later, as the car sat on the lot, the battery burst into flames. It engulfed the Volt along with three other vehicles parked nearby.

Fox News blamed Obama. Neil Cavuto, a Fox commentator, said the Volt was part of a gigantic social disaster that would lead to divorces "when someone forgets to plug it in," not to mention a conspiracy. "Someone bought off *Motor Trend* to say it was car of the year," Cavuto said. "You have to be a dolt to buy a Volt." The vehicle had nothing to do with Obama and in fact was conceived during the George W. Bush administration. But by embracing electrics, Obama infuriated the right.

The carping grew when two more fires occurred during tests just six months later.

The thing about large lithium-ion battery packs was that if you were not going to use them for a long time, you were advised to drain them of electricity. When fully charged, they could be unstable. In the case of the first fire, the ramming pole had punctured the battery pack, resulting in a leak of coolant, a short circuit, and ultimately the fire. It was not a novel occurrence—cars often exploded during the crash tests, especially when they carried a tank full of gasoline. You had to destroy something in order to discover how safe it could be. But nuance was lost in the news coverage. As oil is associated in the public mind with crude-soaked seals, the Volt seemed doomed to correlation with fire. The fires threatened to sink GM's flagship electric.

But then the carmaker defused the crisis atmosphere with a wily combination of candor—spokespeople came out quickly with as many details as they knew—and a spirited campaign of customer service: company representatives called every Volt owner in the country personally—some 6,100 of them—with an offer to buy back their car, replace their leased vehicle, or temporarily lend them a GM model of their choice. A few dozen took up the offer. But mainly they didn't. The image was of a proactive company that believed in its product and was taking charge of its reputation.

It was about this time that Chamberlain was shopping for a new car. He and his wife Kathy liked their old Honda sedan and favored buying another—their son Fletcher was about to turn sixteen and would be given the old car. The new vehicle would be Chamberlain's. Considering that NMC was in the Volt, it would not have taken much to prod him at least to take a look at it, but the math caught his and Kathy's eye: they were spending about $240 a month on gasoline and the Volt would cut that down, perhaps to zero. They still planned to buy another Honda but decided to stop by GM, too.

The GM man asked Chamberlain a couple of questions. Then he tossed him the keys.

"Why don't you drive it for a few days?" the salesman said. "Take it home."

"Seriously?"

"Yeah, I just need a copy of your license."

The salesman had required only a sentence or two to read him. Chamberlain returned three days later.

"You did a terrible thing," he said. "Now I want one."

Chamberlain said that it wasn't only his personal connection to the car that decided him. Notwithstanding the opinion of Fox

News, he agreed with the assessment of *Motor Trend*, which was that the Volt was "a game-changer." The Volt was the future, he said, "something that is amazing."

He chose jet black.

The evening arrived to pick up his new car. He haggled a final time with the salesmen over whether to lease or buy, especially since, over time, the battery would certainly weaken. At last he said he would buy. He thanked the voluble salesman, Jim Vegetabile. "Well, I told you I was gonna take care of ya," Vegetabile said. "See, some people don't believe you when you have a different kind of last name. They don't know it's real."

Chamberlain climbed into the Volt. A smile crossed his face. "I've never had a payment this big on a car in my life." He began to fiddle with the LCD display.

"Listen to that music. We need to turn this off. How do we turn that off? There we go. Listen to this. Oh God, this is sweet. All right, now how do I get out of here?"

He was nervous.

"I'm damn excited," Chamberlain said. He pushed the starter button and moved quietly into the night.

Chamberlain had calculated that if he did use a lot less gasoline for the commute to Argonne, he would, over the life of the car, save enough to justify the monthly payments to which he was now committed. One wondered whether the car was actually more a fashion than an economic statement.

"I'm not into that," he said. "Honestly, it's a toy."

The LCD told him his foot was not on the pedal—he was gliding. While he was doing so, the car was converting energy from the wheels to charge the battery. Chamberlain was watching all this but he was also late to pick up his daughter Abby, who was rehearsing for a Christmas concert at school.

Eyeing a green ball central to the display, Chamberlain saw that he was at "Level 1" energy efficiency, which was good. A friend had told him that the Volt originally accelerated so quickly that GM installed a governor to slow it down. That was because, unlike conventional vehicles, electrics have no gears. "They have to really kind of chill that out or else people would be killing themselves," Chamberlain said.

One feature that impressed him was that GM had designed the Volt with flexibility to be equipped with any propulsion system. A later iteration could be fitted with natural gas, ethanol, hydrogen, or any other technology. That was important, because no one knew which direction fossil fuels were moving.

The green ball was far from the optimal center—Chamberlain had failed to discipline his acceleration and braking. His pedal style was inefficient, the car was telling him. The LCD said that he had 31 miles left on the battery and 269 miles in the gas tank.

He reached Abby's school. Chamberlain tried to connect to the Volt's Bluetooth software while he was waiting. He couldn't figure out how. He eventually found the correct button, but the battery was quickly discharging, drained by the seat-heating mechanism. Its capacity was still limited.

In the subsequent days, Chamberlain experienced seductive quiet. At once, daily neighborhood noise—lawnmowers, telephone repair vehicles, semi-trailers, and rushing traffic—became much more conspicuous, and much less wanted. The car rode smoothly and attracted shouted, admiring comments from passing cars. At night, Chamberlain plugged it in at home. During the day, he charged it at Argonne. The Volt was sanctuary.

25

A Chance to Win the Lottery

To win the Hub, Argonne had to convincingly explain how it would invent the battery that finally challenged the energy density of gasoline. To get there, it had pulled together many of the best battery scientists from across the country—Venkat Srinivasan from Berkeley, Yet-Ming Chiang from MIT, and Yi Cui from Stanford, not to mention Thackeray and Amine, plus the chief technology officers of companies like Dow Chemical, the lithium-ion battery company A123, and Johnson Controls, one of the world's largest makers of lead-acid batteries. But in directing the team, Chamberlain had the tricky task now of negotiating how to divide the anticipated spoils of blockbuster inventions. If the partners were not confident that their contribution to breakthroughs would be fairly credited and compensated, they were unlikely to put in the colossal effort that would be required to produce the better battery.

A fight broke out between Argonne's lawyers and its industry partners on one side and battery inventors on the other. Chamberlain was in between. He was prepared to grant the companies first rights to any inventions. But the reward to the inventors had to be specified, too. In a meeting with company executives, Chamberlain said the scientists were accustomed to public sponsorship, embedded in which was the incentive of royalties. It was like the lottery—they had no sense of entitlement to royalties *per se*, but they would not part with at least the *potential* of an outsized payday. Yet that unassailable aspect to lab work would vanish with a change from public to private sponsorship. That was because to industry, the calculus of success was measured by *company* triumph. In the case of a venture such as the Hub, Dow, Johnson Controls, and the rest would be seeking a particular profit formula that involved *certainty*. It would allow them to calculate how much they would need to charge for a product five to ten years before its actual manufacture and thus assess in advance the wisdom of a given investment. That this was a rare moment when they were petitioning Argonne for its help did not curb their demands. If Chamberlain could provide that certainty, the companies could live with a formula that satisfied the inventors as well.

"The inventors have an implicit conflict of interest," a lawyer told Chamberlain. "They should not be involved in any way or shape or form in any kind of decision."

The assertion that only the scientists, and not the companies, were self-interested was odd. It reflected a contradiction. The scientists were paid to research for the good of the nation and could gain personally in the remote case that an invention struck gold. But the relationship was symbiotic—American companies were increasingly dependent on the national lab system. They had a stake in keeping the inventors happy.

Over the years, blue-chip companies had become proud glorifiers of tinkering. Silicon Valley no doubt thought it was living up to its reputation for inventiveness, but a more beautiful and useful smart phone or tablet wasn't equivalent to the invention of the light bulb. Excluding its universities, the Valley was not producing Nobel Prize–level work, something that Bell Labs accomplished seven times.

The Valley had yet to concede the hollow place it had become, but industrial companies had—they knew they lacked a crucial dimension: they could no longer hope to make the big leap on their own terms. That recognition underlay Dow Chemical's eagerness to join En-Caesar or the Hub. It hoped that, in collaboration with Argonne, it would invent and commercialize the next revolutionary product.

Argonne just as eagerly embraced this company interest, which could be a source of scarce research dollars, independence, peer respect, and validation. But did that mean, as the lawyers argued, that the benefits ought to be tipped in the companies' favor? That they should play no role in determining their own compensation?

"They have no say in it at all," the Argonne lawyer replied.

"Except they can say if they want to work or not," Chamberlain said.

"And you say, 'Do you want a job or not?'" the lawyer said.

"Can I? I can't force them to do work."

"If they don't do the work, then you put that on their evaluation and use the evaluation to push them out," the lawyer said.

Chamberlain stared impassively. "That is terribly unrealistic for a star scientist who already has a hundred and fifty percent more work than he can handle," he said. "We just have to figure out a way to properly incent employees without creating some kind of nasty competition."

"They won't know what the terms of the contract are."

"That's true," Chamberlain said. "They're not supposed to."

Ultimately they would settle on a formula providing a third of the money to the institution where the invention was created; each Hub member would decide for itself how much to cut to the inventing scientist.

Chamberlain had been alone in protecting the scientists' backs. He bent to the assertion that the scientists were subordinate in the pecking order but secured their right to fight within their labs—where each probably exerted the most influence—for just IP compensation.

He focused on the big picture, which he said was national and geopolitical. He wanted the United States to prevail in the battery race. To do so, the United States had to develop a manufacturing supply chain, a necessity if you were thinking of dominating an industry. That meant licensing IP and supporting capable start-ups like Envia, in addition to creating a new invention system to help remake the commercial capabilities of incumbent giants like GM and Dow Chemical.

26

"There Is a Problem with Your Material"

Until Sujeet Kumar paid a visit to Argonne in 2007, the NMC in a way did not exist. It was only when he began pressing for a license, congratulating Thackeray for creating what he called the only cathode with a chance to challenge oil, that the South African—that anyone at the lab—grasped how exceptional the invention was.

It was the nature of basic science labs. When Goodenough and Thackeray made their early back-to-back achievements—the former's lithium-cobalt-oxide in 1980 and the latter's spinel a year later—there was no market for their advances. Rechargeable lithium-ion batteries became commercial products only a decade later. When Sony commercialized Goodenough's battery in 1991, it became the go-to formulation for virtually every laptop, smart phone, recorder, or really any battery-enabled consumer device. Goodenough's batteries lasted longer than the technology they superseded—nickel metal hydride—and did not suffer nearly the severity of capacity loss after long use. Even two decades later, lithium-cobalt-oxide batteries remained the world's workhorse consumer battery.

But all of that was later. In contemporaneous time, the drive behind both Goodenough's and Thackeray's breakthroughs was little more than "academic curiosity." No company showed the slightest interest in their inventions. Not that they never contemplated commercial application—they did; in strolls and hallway conversations on the Oxford campus, they shared their hopes for market adaptation. But the personal payoff became a citation in others' papers and continued funding.

The inspiration to use lithium-ion to revive electric cars, though, came later still. Lithium-cobalt-oxide was too expensive—specifically the ingredient cobalt—for serious contemplation in passenger vehicles. It packed a wallop of energy density—the best among any commercial battery—but was economically feasible only for compact purposes, meaning small electronic devices. When Toyota pioneered the modern-day push into electrics in Japan in 1997, its Prius hybrid again contained nickel-metal-hydride batteries.

No other lithium-ion formulation was ready, either, well into the 2000s. Thackeray's spinel bested lithium-cobalt-oxide in the laboratory, but in actual use it quickly degraded; it would be the mid-2000s before the degradation problem was resolved and spinel could be used in the 2011 Volt.

It was the same with Thackeray's NMC. He and Chris Johnson were well on their way to their breakthrough invention by the time of the first Prius, but they were in no rush because they knew of no interest in lithium-ion for automobiles. Again, the motivation was personal inquisitiveness.

After their flush of triumph on patenting the NMC first, they were followed by Dalhousie's Jeff Dahn and the New Zealanders. Colleagues congratulated Thackeray and Johnson; peer recognition grew. Even then Thackeray did not sense a world-beating breakthrough; he told Argonne's intellectual property team to seek an American patent, but not to bother with the expense of an extraordinary protective effort. Do not file for international patents, he advised Argonne's licensing team. Having missed out on the original U.S. patent, 3M, possessing the rights to Dahn's version of the NMC, filed for international rights.

But NMC 2.0—the tweak on the original composition—excited a stir in the battery community. It was identified as the next great battery material, superior to another futurist rage, lithium-iron-phosphate, a commercial version of which was created by A123, the celebrity battery start-up. A123's chemistry—built on Good-enough's original breakthrough—dispensed a burst of energy that was perfect for power tools. But because it worked at an average of just 3.3 volts, it underperformed when it came to endurance—the ability to stay charged for long distances, a quality sought by car-makers. Until now, automakers appeared prepared to adopt it anyway. But NMC 2.0, operating at higher voltage, provided im-proved endurance, threatening A123's perch.

A123 was not easily ceding the top spot. Bart Riley, its muscle-bound chief technology officer, had obtained samples of NMC 2.0. If it truly was revolutionary, Riley was prepared to make peace and figure out how to adapt to the new market. But he told his re-search team to run it through all the tests.

From the company's Waltham, Massachusetts, laboratory, Ri-ley received an e-mail from a forty-one-year-old South Korean staff researcher named Young-Il Jang. NMC 2.0, Young said, ap-peared to have a problem. And not just any problem, but one so substantial as to possibly doom it outright for use in cars.

Young told Riley and other colleagues copied in the e-mail that the jolt of voltage that gave NMC 2.0 its potency also seemed to thermodynamically change it. When the high voltage forced much of the lithium to begin shuttling, thus removing the cathode's pil-lars, the structure sought to shore itself up and keep its shape. Other atoms rearranged themselves. Nickel took the place of lithium, and cobalt of oxygen. When the lithium returned, its old places were occupied. It had to try to find a new home. Thermody-namics made the atoms seek a new natural balance. The voltage steadily declined.

Hence in actual application in an automobile, NMC 2.0 might not provide the consistent potency suggested when Thackeray was working on coin-size test cells in the laboratory. Unless the atomic reorganization could be controlled, Young concluded, the material might never find use in a car, which required reliability. In a gasoline-driven vehicle, the driver expected the engine to deliver more or less the same propulsion each time the accelerator was depressed—the pistons had to push out a smooth flow of power continuously, every time. It could not deliver the acceleration of a Ferrari the first day and a Mini Cooper on the hundredth. Similarly, in an electric system, the voltage in the second cycle could not differ from that of the fiftieth; you could not create a dependable, ten-year propulsion system with such instability.

While perhaps bad news for Thackeray and Sujeet Kumar, Young's find enthused Riley. He had heard plenty of passionate discussion of NMC 2.0, but not a word about this potential problem, what came to be called "voltage fade." In Riley's view, that created a substantial opportunity for A123. Over the years, no private battery company had seemed as bold, shrewd, and quick as A123. In this case, if Young was in fact the first researcher to notice voltage fade, he—and A123—could possibly be the first to solve it. A123 could patent the answer and then reap the profits of enabling NMC 2.0 for the electric-car world.

Riley put in a stream of calls, chiefly to Jeff Chamberlain and Sujeet Kumar. To each he said the same: What were they doing at the moment to solve voltage fade? "Let me check," Chamberlain said. Kumar answered similarly. Clearly neither was on top of the problem.

At a battery conference in Montreal, Riley ran into Thackeray. "You know there's a problem with your material," Riley said. "It could be a doom factor."

"It's something we have to address," Thackeray said. In fact, the voltage question had recently been raised in the Battery Department, but until running into Riley, Thackeray did not consider it important.

Back at Argonne, Thackeray checked with Sun-Ho Kang. A few months earlier, Kang, the South Korean who later would go to Samsung, had noticed unsteady voltage in NMC 2.0. Like his counterpart at A123, he thought that if voltage fade could not be stopped, you could not use the NMC 2.0 in a car. The Volt did not suffer from the problem. It used the original NMC, which did not undergo the same supercharging; a maximum of 4.3 volts was applied to it, lower than the 4.5 volts that activated NMC 2.0's atomic-level chaos.

Riley was suggesting that the parade of companies that had paid to license NMC 2.0—not just Envia, but BASF, GM, LG, and Toda—were holding a seriously flawed product. As his researcher had stated, NMC 2.0 perhaps could not be deployed for the purpose for which it had been purchased—longer-range, cheaper electrified vehicles. At least in its current state, it perhaps could only be used at lesser voltages, which would mean performance not much different from the lithium-cobalt-oxide batteries commercialized two decades before. There might be no reason for anyone to absorb the expense of switching to NMC 2.0.

If you asked the battery guys at what stage they understood that there was a problem with NMC 2.0, it prompted a nervous response. They would go quiet, glance around, and provide not quite precise answers. This conveyed the impression that either no one knew the precise answer or no one wanted to disclose it. The reason being that, if you looked at the situation squarely, you could not escape the conclusion that Argonne had in fact sold the companies a faulty invention. Not that the companies themselves were off the hook— the engineers, venture capitalists, and other executives and staff who had signed off on the licenses had to be in some hot water among their bosses, too. If anyone was predominantly responsible, it was the Thackeray team, because their names were on the patent.

Chamberlain, who had led the negotiations on Argonne's behalf, said simply, "We didn't know about it." But how was that possible? "Because making a product is not the scientists' objective. You have to look at a certain data set to notice the fade," he said. "If you look at a different data set where all of your requirements are for capacity, you can actually miss the voltage curves." He added, "That is why interaction with industry is so important, because if you are making a product, like a battery that is going into a car, you look at everything like this."

What Chamberlain described was precisely why Riley's team might manage to steal a march on the industry. The reason why no one had picked up on voltage fade was that it had been a problem in no other major battery chemistry. So when battery guys evaluated this new chemistry, they skipped the voltage. "There was a blind spot," Riley said.

Papers Thackeray and Kang had published as far back as 2007 revealed weaknesses of NMC 2.0. The papers included charts revealing the fade phenomenon. Thackeray argued that that meant that Riley had stumbled over nothing new. But he was wrong. Thackeray and Kang published charts of voltage fade without explaining its significance, because they themselves did not grasp that it was a potential showstopper for the NMC 2.0. Riley did.

As for why Young-Il Jang did check for it, Kang said he himself tipped off his A123 counterpart. But if so, Young-Il was already prepared to find the fade because he had observed a similar phenomenon in a cathode on which he worked for his doctoral thesis in the late 1990s. That allowed Young and Riley to move faster to report the finding.

Whatever the case, Riley, armed with what he called a major scientific "scoop," now pursued talks with both Chamberlain and Kumar. With the latter, Riley held out a seductive offer—A123 and Envia could embark on a year-long joint research project on voltage fade. If it succeeded—if they solved voltage fade—A123 would buy Envia for $120 million.

Kumar conveyed the offer to his board, which said that Riley was essentially attempting to steal Envia. If Envia solved voltage fade, the board said, it would not require any relationship with A123— Riley was assuming no risk at all with his offer. Kumar instead should himself go into battle and attempt to solve voltage fade.

Around this time, Kang was driving to Yellowstone National Park when his cell phone rang. A friend delivered tragic news. Young-Il Jang had died. Just forty-two, he had suffered a freak heart attack in his sleep.

Absent Young's intellectual lead, A123 soon became distracted by other matters, including its viability as a company. Sales were not rising as fast as the founders expected, and a solution to voltage fade dropped from Riley's agenda.

Few of the battery guys agreed with Riley and Kang's assessment. And even if they *were* right, the market for such powerful batteries was embryonic—it seemed unlikely to soon require battery packs of the size that worried the A123 men. No one in the industry was demanding anything different for the moment either. So no one was in a rush to look deeper at the chemistry.

But then the industry abruptly changed its mind. GM, Nissan, Ford, and carmakers around the world began to seek batteries that could do much more than boost the mileage of a Prius. They decided that NMC 2.0 stood the best chance of providing such capacity.

Kang and his young colleague Kevin Gallagher were convinced that voltage fade was an existential problem for NMC 2.0. Unless it could be overcome, the composition was dead as far as use in an automobile. Over several months, they tried but failed to isolate the problem; all they managed to discover were conditions under which it became more severe.

In October 2010, a big battery conference was to be held in Las Vegas. Until now, only Argonne, Envia, and A123 seemed to be talking about the potential problem of voltage fade. Kang told Gallagher that he ought to disclose it publicly. Gallagher was nervous. Who was he to "rock the boat" on NMC 2.0? But Kang said, "I feel a responsibility. If we know the problem, we have to let it be known so people will work on it. And we should work on it, too."

Gallagher delivered the talk. The high-capacity material was still terrific, he said. Except that when you activated the NMC by applying more than 4.5 volts, you triggered fade. He held back from Riley's apocalyptic forecast but had put the industry on notice that NMC 2.0 had a defect.

Not long afterward, Kang announced his move to Samsung. Department of Energy staff summoned him to Washington. They wanted to hear more about voltage fade. A few days before his departure to Seoul, Kang sat before six Department of Energy officials with his slide deck. His core message resembled A123's: NMC 2.0 required a fundamental fix.

How did some of the best minds in batteries overlook a defect this basic? Voltage fade was deeply pernicious, Kang said. It was what Chamberlain said—if you were employing the standard measuring tools, determining a battery's stability by checking its capacity, you would notice nothing wrong with the NMC 2.0. From cycle to cycle, you observed a stable composition. That is what Thackeray and Johnson saw and reported in their invention. Voltage fade became conspicuous only when you incorporated gauges of stability that, while familiar in industry, were highly uncommon in research labs. Only then did you understand that NMC 2.0 was profoundly flawed.

Peter Faguy, a bearded and garrulous Department of Energy official, dropped in on the battery guys. He was one of the officials who had met with Sun-Ho Kang. Now he sat with Chamberlain and Thackeray in a small conference room at Argonne.

"You've gotta put a big team on this," Faguy said. He meant voltage fade. "We've got to sort this out in three or six months." There was no time to waste.

The Argonne guys stared at Faguy. They agreed more or less with his assessment, but the deadline he suggested was "laughable." "It just didn't seem to be realistic," Thackeray said later. They did not challenge him, but once he was gone, they quickly forgot about the directive.

Considering the egregiousness of the flaw, the battery guys—at least on the surface—seemed oddly relaxed about it. It was the old divide—once an invention was out of the laboratory, especially if it was already licensed, the scientists were on to the next project. Thackeray appeared unconcerned about the reputation of his signature material. He showed no sign of turmoil as he worked with a postdoctoral assistant to try to fix the fade.

Months passed. Chamberlain and Tony Burrell, his new deputy, were summoned to Washington.

"Solve this friggin' problem," said David Howell, Faguy's boss and head of battery development at the Department of Energy.

Chamberlain said he had not been entirely ignoring Faguy's prior instructions. He had a plan to attack voltage fade, he said. He would put together a team at Argonne and its members would concentrate solely on the problem with NMC 2.0. It would cost a few million dollars, split between the Department of Energy and one or more of the licensing companies. He believed it stood a good chance of working.

"You have plenty of money in your budget," Howell said. "Use it." He seemed exasperated. Chamberlain should repurpose his budget into a program that made good on NMC 2.0 by solving voltage fade. He was not allocating another dime.

Later, speaking privately with Howell, Chamberlain asked whether he had been pressured by the companies to come down so hard. Howell said that, no, he hadn't. Chamberlain saw that Howell was worried about the prospect of corporate disenchantment. He wanted to avert an angry call from GM or anyone else. Given the administration's agenda and the hoopla over the NMC licenses, Howell might not have been surprised if the president himself rang.

If Argonne solved the problem, there would never be such calls. If there ever were, the next best thing to a solution would be to say that he had dispatched all his troops on a monumental effort to find an answer.

A month later, Burrell called Howell with the Argonne plan. The lab would reshape the entire Battery Department around the effort. Almost every researcher in the department would be reassigned to the problem, and other specialists brought in from other parts of the lab as well. They would plumb why voltage fade was happening and seek a solution. The effort would cost $4 million a year, all of it taken from existing Argonne programs funded by the Department of Energy. Argonne anticipated a slog—the answer would not come in weeks or even months. It could take three or four *years* of work. The effort might not succeed, and if failure seemed likely, they would pull the plug early.

Faguy was not satisfied. You were building a house of bricks and you did not lay down a new brick until the previous one was settled in place. Voltage fade was a loose brick in the edifice of the electric car. He thought that NMC 2.0 could underpin big new industries in an otherwise laggard American economy. Car companies around the world felt it had the best energy density, capacity, and voltage of any material currently within reach. "This is the best chance that we have of supporting U.S. industry for the next five years," Faguy said. But only if it actually worked as advertised. The Department of Energy—Argonne—"has an obligation" to fix it. He tried to sound undemanding, but Burrell's potential four-year time line was too long. He said, "We don't have an expectation that it will be solved in the next few months," but "we also certainly don't expect to put money into it in the next three years." The middle ground was eighteen months—by then he wanted either a solution or, at the very least, a description of "the foundation of a solution."

Voltage fade was now one of the field's defining challenges.

27

An Engineering Solution

The wall of Envia's conference room was decorated with a photograph of piles of lithium carbonate in a field, white pyramids of powder that resembled cocaine. A plaque marked the 2009 collaboration between Envia and Argonne on the NMC. In a corner were Envia shoulder bags and on the table a selection of the company's cylindrical and flat batteries. On the wall, a plaque denoted "Envia's Values," which included "problem solving, not just program management" and "develop products, not just patents."

Kumar stood with three scientists. "This is the core R&D group," he said. Chad Bowling, a blond and serious Louisville, Kentucky, native, had graduated from Stanford with his bachelor's in chemical engineering two years earlier, in 2009. Stanford engineering graduates were typically snapped up by the high-tech industry, but with the economy in recession, Bowling couldn't find a job. He was subsisting on dinners of bean burritos and had started applying to restaurants to wait tables. Then a former professor recommended him to Kumar. Envia was seeking to hire a Ph.D. scientist. After speaking with Bowling, Kumar agreed to try him out as an intern. Six months later, he promoted Bowling to full-time technician.

"We call him 'Doctor Chad,'" Kumar said. Normally one required a doctorate for the work that Bowling was carrying out, but "Chad is getting his Ph.D. at the Envia School."

"Right, exactly," Bowling said. His voice was firm, his manner still.

Bowling worked under Mani Venkatachalam, a physicist and materials scientist from the southern Indian region of Tamil Nadu. Venkatachalam was Kumar's chief electrochemist. Kumar directed him to "go out and be ahead of everybody else—get the highest capacity, the highest power, maximize the voltage. Don't worry about the customers. Do what you want to do."

One of their aims was to navigate voltage fade. Like everyone else, Kumar had a blind spot until Riley called. The A123 man was right—it needed to be brought under control. Kumar was secretive about Envia's precise approach to fade—how precisely Bowling and Venkatachalam, the company's voltage fade team, were progressing. You could see that Bowling would make electrodes, laminates, and coin-size test cells and take the results to Venkatachalam and the rest of the research group, whose members would remark on his data. They would speak of "nano-coatings" and an element or two of which they were made. But because of the value should they succeed, none would discuss details such as the specific quantities of the ingredients that went into the concoction. No one working on the problem anywhere was speaking that openly of their work.

Some two dozen men filed into the white, spare conference room. All toted laptops, which they put down, open, on the long rectangular table, and sat. Kumar handed over the meeting to Bowling.

"I just got these numbers like an hour ago," Bowling said, displaying his slide deck on the screen. "With the normal process, I mixed for five hours, then put on an oxide coating where we would mix two hours, over seven hours in all. Then do it again for the next composition."

"So this is not fired?" one of the scientists asked.

"No, it just aims to establish one layer, and then another layer," Bowling replied.

A compound doped with aluminum curtailed voltage fade best, he said. The question was in what proportions and at what thickness. He was using a scanning electron microscope to scrutinize the aluminum impurities and determine their optimal size.

"The ten-K magnification makes it pretty clear," he said.

"Nice fibers," Kumar replied.

Bowling flipped through more slides and revealed that when he applied thicker layers of the coating to the electrode, voltage fade seemed almost to flatline. Kumar's voice revealed increased excitement. "This is kind of a breakthrough," he said.

Bowling replied evenly, "This is definitely the most drastic of all the average voltage improvements. And it's showing no sign of a top in terms of improvement. I mean, as high as we've gone, it's continuously improving."

"We can go more," Kumar said. He sat back in his seat. Everyone was silent.

"The whole world is here," Kumar said, pointing to one section of the latest slide. "They have more than ten percent fade. Now we have solved the issue by at least eight percent. And we're going to solve the last two percent, right? The last two or three percent?"

"The last one percent now," someone said.

"That's good. It's really nice," Kumar said. "So we are very, very close to solving this problem."

After the researchers filed out, Kumar said, "Chad has one of the best experimental hands I have seen. There are lots of top scientists trying to solve this problem. We're ahead of everybody." He said that once one of the samples showed an absence of voltage fade, "we will know we have solved the problem, and we can go back and say, 'Okay, what *really* solved the problem?' and then science is done."

He meant that you could tinker in order to reach a favorable outcome, but tinkering didn't explain why the atoms were behaving a different way. When you went further and discovered the reason, you added to the field. The answer probably did not matter to any of their customers, but it was important to be able to explain the result, if only to themselves.

This was Kumar seeking an engineered solution to voltage fade. For a start-up, and really any private company, there often was not the time to figure out the atomics of a problem and root it out—what Argonne was attempting. You instead used your instincts to circumvent a difficulty and get a system to work as well as possible. The result would fall short of ideal. But it would be sufficient to start to sell Envia's version of NMC 2.0.

Tempers were fraying over NMC 2.0. Envia had won contracts to collaborate with Argonne and Berkeley along with a couple of universities on the battery challenges, and the Department of Energy was closely watching its adaptations of NMC 2.0. But Kumar also had to carefully compartmentalize his work so as not to compromise the deal with GM or the other carmakers with whom he was conducting research. A few months earlier, a Department of Energy official had remarked to Kumar, "You guys should bring a solution in sooner rather than later. You owe us a response." Kumar had responded, "My guys are already going fast. I want them to work without any more pressure." Then he had told his men, "Don't worry about any deadline, any time line. Do whatever you feel like. Let's try this crazy idea, that crazy idea." Envia was a start-up, not a national lab. The priorities differed.

Kumar thought back a year earlier, when he first began attacking the problem. His team hashed over the different theories as to why the material was unstable when you juiced the voltage. Was the change happening deep within the material, or on the surface? If the change was in the interior, you could dope the material to try to trigger a different reaction. If it was on the surface, you could paint on a coating to try to alter how it reacted in contact with the electrolyte. Initial data pointed to the surface, so the team started a list of every possible variety of nano-coating that might create stability. Some of the coatings would inadvertently prevent the lithium from shuttling properly. But then Bowling and Venkatachalam combined two different coatings and found that, regardless how thick they painted the cathode, the lithium moved just fine.

Kumar understood that the answer was a composite of coatings.

But he still didn't know what the composite was arresting or why it succeeded in doing so. He didn't have the instruments that could figure it out. Lawrence Berkeley National Laboratory, twenty minutes north of Envia, had a powerful transmission electron microscope that could peer deeply into the material at high resolution. The lab charged $400 an hour to use it. Kumar would be studying numerous samples, at $5,000 to $10,000 each. He imagined the bill might run over $100,000, too much merely to "do science," he said.

Alternatively, he could write a proposal to persuade LBL to do the work out of its federal funding. But in that case, Kumar would have to publish his findings. That was the law—keeping the results secret would be unfair because the public was paying for the work.

Since Kumar sought to keep Envia's intellectual know-how confidential, the best solution was to pay. But that violated Kumar's and Kapadia's tenets of hoarding cash.

Kumar approached a friend who worked at LBL. "I want to bring you some of my material, to analyze it, but take all the material back and keep my IP," Kumar said, adding that he did not want to pay. "How can I do that?"

He couldn't.

Though they said otherwise, Thackeray and his team at Argonne seemed bothered that Envia—a start-up licensee—might find the solution to the voltage fade conundrum first. The industry talk was that Envia was on the verge of an answer. Outwardly, all of them said they wished Envia well. Its success would be the same as Argonne itself triumphing. But an Envia victory could be viewed as a mark against Argonne, a sign that it couldn't clean up its own mess.

Faguy thought Envia was not alone in preparing to settle for an engineering solution. Short of actually solving the fade, the other NMC commercial licensees also seemed to be looking for a work-around that made the material usable. Even if it didn't perform as promised, it would still improve on current chemistries. Having paid for the NMC, they would find a use for it in smart phones and other electronics. At that stage, "the case will be closed for them," he said.

But as far as he and Howell were concerned, that wasn't good enough. The Department of Energy wanted "a bedrock understanding of what the hell is going on" within the cathode, along with "the whole gamut of possible solutions."

Further in the future, Faguy saw the problem as a dress rehearsal for nightmares to come. The battery race would involve a series of unforeseen, terrible problems that you simply could not recognize in the tiny volumes and coin cells produced in the national labs. You needed a ton of the material and hundreds of cells, and you had to charge and recharge them again and again before the problems surfaced. Only then could you think about the solutions necessary to get the technology into a car. Voltage fade was a test run for how the Department of Energy—how Argonne— would resolve the crises to come.

28

Going Deep on the Fade

T hackeray went to California to see what precisely Kumar was up to, taking along his postdoctoral assistant, Jason Croy. Just how close was Kumar to solving voltage fade?

Croy, a slight, thirty-seven-year-old physicist with cropped blond hair and a frequent smile, had finished graduate school just a year earlier. He grew up in Frankton, an Indiana town of 1,800 people, and came late to physics. For nine years after high school, he, his older brother Johnny, and three friends toured the Midwest in a heavy-metal band called Connecticut Yankee. Croy sang lead and Johnny played bass on two records of original music, including a song called "$F=G(M_1M_2/D^2)$," Newton's equation for gravity. Science was always there. He built a telescope and took an astronomy course at Ball State. One day, a lecturer described the possible use of plasma physics to understand nuclear fusion, the process that powers the sun in which two nuclei combine, releasing energy, and Croy "thought it was the greatest thing I had ever seen." He enrolled full-time and went on to earn a Ph.D. from the University of Central Florida. Croy's aim was "something that would really have an impact." In 2011, Thackeray advertised a postdoctoral position on Argonne's Advanced Photon Source, the 3,600-foot electromagnetic loop that produces intense X-rays that researchers used to examine the materials they were creating. They called it the "beam line." When they were working on the beam line, they called it "beam time." It was a highly prized tool, and a coveted job working with it, because of the deep images it produced of the atomic structures of their work. Though a decade older than most other applicants, Croy won the position.

Croy's years on the road left him uninhibited and quietly commanding. Thackeray was a powerful public speaker, but after watching Croy's delivery a couple of times, he seemed almost to prefer that the younger man present their findings. In the Envia conference room, Kumar introduced the Argonne men to his assembled researchers and said that without Thackeray, Envia would not exist. The Envia guys were visibly elated to be near the inventor of the NMC. Now they all turned to Croy, who was standing. Croy said the slides assumed two ways to understand voltage fade: it was either repairable or forever unmanageable, the latter because of the immutable laws of thermodynamics, the most basic physics of energy. The answer, he said, was actually both—voltage fade challenged the limits of fundamental physics, but there could be a fix. To get there, he and Thackeray had used the beam line to explore the bowels of the NMC.

They observed that the nickel and manganese had wanderlust. The metals liked to move around through the layers. It was their nature—once the lithium shuttled to the anode, taking a bit of oxygen out of the cathode, the nickel and manganese could not help but shift in order to find a new, comfortable balance. By the time the metals settled down, the material itself was changed—its voltage profile was vastly different. For a carmaker, such a transformation was unacceptable. But how could you stop it?

One of Thackeray's solutions was elegant—to accept the predilection. He proposed that researchers stop trying to evade physical laws and instead work with them: Croy artificially created NMC 2.0 with the manganese and nickel already in postactivation balance. The outcome was a cathode that was less energetic—it might deliver 190 watt-hours per kilogram, or a little over two thirds of the original estimate. But it was also free of voltage chaos. It was stable and usable, which was an improvement.

Croy explained to the Envia group that the method was first to treat the Li_2MnO_3 in acid, water, and nickel nitrate. That leached out the lithium and some oxygen. The action jumbled the cathode's lattice—the nickel was now not where chemistry said it should be. But that did not matter much at the moment. Croy allowed the water to evaporate. He held on to the lithium he had removed and heated the components. Then he slowly cooled them. The result was a reconstituted and stable amalgam of manganese and nickel in a layered-layered composite structure. The atoms of nickel were intended to serve as props to hold up the structure while lithium shuttled back and forth. He and Thackeray had already applied for a patent describing this new NMC 2.0 manufacturing method. In two weeks, they would amend the application with yet new findings and claims.

Thackeray's aim in allowing Croy to make the private presentation was to pick up any details they could—by way of facts disclosed or questions asked—on Envia's own progress. The South African—like most of the Argonne guys—was anxious to know what was going on in the Newark lab with the NMC 2.0.

For Kumar, too, there was advantage in sharing ideas with Thackeray, who knew the material better than anyone. Until now, Kumar, while not disputing the possibility of a problem of thermodynamics, had adhered to his engineering approach, disregarding the physics and simply trying to get the NMC 2.0 to work. Now, sitting before Thackeray, Kumar recognized that his team had advanced about as far as it probably could. His team was eradicating the relatively minute fade that occurred in the NMC only *prior* to its activization—they were not taking it past 4.5 volts. But some problems simply could not be engineered away—in order to attack what happened to the material after activization, Kumar and his guys needed to start to understand the physics.

In the subsequent discussion, Kumar said he wanted to collaborate with the Argonne men, and Thackeray felt the same. Envia and Khal Amine had worked together before, but the project was not as commercially sensitive as NMC 2.0. The trouble now was the IP. Kumar worried that, if he teamed with Thackeray and

Croy, whatever learning emerged might leak to Argonne's other NMC licensees, which could jeopardize his business advantage in collaborating in the first place. For Thackeray, the concern was precisely the opposite—how could he possibly work with Kumar and also produce new public knowledge?

They spoke for more than three hours. Thackeray and Croy had to catch a plane. Thackeray decided that when he returned to Chicago he would speak to Chamberlain about how to work with Envia while protecting the interests of both sides. But the Argonne men left feeling less apprehensive about Kumar. Neither thought Envia had broken through on voltage fade.

When Thackeray and Croy spoke of artificially creating a balanced version of the NMC 2.0 in advance of the fade, they had a picture in mind of its atomic-scale appearance. Thackeray reproduced the image as a drawing in a published paper—a flower pattern of evenly distributed pockets of manganese and nickel. The extra manganese in NMC 2.0—the Li_2MnO_3—that was largely responsible for the battery's exceptional performance also contributed to its instability. The manganese settled down and stopped rattling the structure when near nickel. So wherever you had manganese, you wanted to make sure nickel was also present. The flower pattern represented the best depiction of that balance.

Could you recreate this flower pattern on the nanoscale? Thackeray said such a configuration had never actually been observed, so there was no way to know if it could be constructed. Croy was slightly more confident. He thought he and Thackeray could create a nickel-manganese flower pattern "to some extent," though some isolated clumps would remain. "It is just very hard to get things exactly like you want and even harder to get them to stay there," Croy said. "It's like your body going to hell as you age. There's just no way out of it! So we work out to slow the decay. We need to find the equivalent for batteries."

Croy stared at his screen. Charge-discharge charts were arrayed across it. Eight more such charts were taped across the wall in his line of vision. He continued to be vexed by the mutability of NMC

2.0. One wondered why he should be. If you applied extreme voltage to any edifice, and if you proceeded to remove a major part of its infrastructure, it might cave in. If in this case you were removing and reinserting stuff again and again into an electrode, *shouldn't* it shift around naturally? That was true, Croy said, but "when we put it back in, we still want it to go back to the way it was."

Thackeray and Croy seemed to have lost their hope of achieving the performance promised for NMC 2.0. They were not likely to attain 280 milliampere-hours per gram, the original goal, but rather a stable 230—if they were lucky. Seeking more amounted to greed. "It's like asking me to be Tiger Woods," Croy said. "I can like golf a lot and practice, but I just won't be him."

Two or three of the battery guys proposed continued battle with the physics. They argued for the introduction of pillars into the NMC 2.0 lattice: nano-size metal atoms, if strategically placed, might hold up the lattice while the lithium was shuttling, they said, preventing the collapsing sensation that sent the nickel and manganese scampering into a rescue position. Croy walked down the hall and into Laboratory X-165 to try out the idea.

Just inside the door, he grabbed and put on a white lab coat, blue gloves, and protective glasses. Then he found a small plastic jar marked "Li_2MnO_3," and a beaker. He placed a flattened square of foil under these items and a spatula and a long slender pipette next to them. He picked up a container of nitric acid and some water and dropped them into the beaker, which he placed carefully into an automatic magnetic stirrer. That done, Croy grabbed another small sheet of aluminum foil and folded each side over to create a margin. He used a thumb to fashion a space in the foil where he could collect crystals of material. The metals might prop up the cathode, but Croy also thought that they could migrate to the cavities created by the shuttling lithium and oxygen. If they thus blocked these cavities, the manganese and nickel might be forced to stay in place and not wander. Either way, he might curtail the fade.

Croy placed the foil on a scale and then scanned a handwritten recipe book. First step—dissolve 0.1927 grams of aluminum nitrate into the solution. He did that. "Now the cobalt," he said, sprinkling in 0.1495 grams of the red crystals and mixing. Green flakes of nickel nitrate were next—1.3609 grams. The solution was now pink. He added two grams of Li_2MnO_3, a rust-colored powder—"kind of like the Martian surface," Croy said. The solution had turned light green. There was enough for a couple of tests. He covered it with foil and put it on the metallic spinner. He would let it spin overnight so that all the chemical reactions would finish. He would dry it in the oven, then grind it. It would go into the furnace, then be ground again and passed through a sieve to ensure that all the particles were below a certain size. Then he would create a 50-micron-thick layered laminate of the resulting solution. All of this would take two to three days.

Croy had carried out this same process previously using atoms of nickel nitrate, but it did not work as well as hoped—the voltage was more stable, indicating that the nickel nitrate did create pillars, but too low. They sought stability at higher voltage. Croy would try again. Unlike the last time, he would place the pillars not in the cathode's metal layers, but amid the lithium. And, rather than nickel nitrate, he would use atoms of aluminum and magnesium. Perhaps they would make the difference.

PART III

RECKONING

29

Orlando

I n February 2012, about a thousand men and women assembled
at an upscale Orlando golf resort called ChampionsGate. There
are two types of battery conferences—scientific gatherings that
attract researchers and technologists attempting to create breakthroughs; and industry events, attended by merchants and sales-
people. Orlando was the latter. A pall hung over the assembled
businesspeople. Americans were not snapping up electric cars:
GM sold just 7,671 Volts the previous year against a forecast of
10,000. There was no reasonable math that got you to the one million electric vehicles that Obama said would be navigating
American roads by 2015, even when you threw in the Japanese-
made Nissan Leaf, of which 9,674 were sold in 2011.

That became even clearer when just 603 Volts sold in January
2012. No one seemed consoled that China was doing even worse,
selling just a combined 8,159 across the country, fewer than half
the American number. Nor especially when, in a conference session, they witnessed the following exchange between a Japanese
presenter and an American salesman:

American: "Do you have any advice for us new entrants into the
business?"

Japanese: "Get a new job."

The Japanese believed the race was already over. They—and
their Prius—had won. Toyota was nearing four million cumulative
hybrid sales worldwide, including 136,463 Priuses in the United

States alone—the world's second-largest car market behind China—in 2011. The Japanese themselves bought 252,000 Priuses. There could eventually be the type of market shift that both Obama and Wan Gang had forecast. But it would not be in the current decade. Until at least the 2020s, electric cars would remain at best a niche product. That would be late for most of the Westerners at ChampionsGate. Unlike the Japanese and the South Koreans, few if any had budgeted for a long struggle. The fever of the prior two or three years began to evaporate.

About this time, ExxonMobil released a fifty-one-page outlook of the world of energy in the year 2040. Such great stabs at the future could not be entirely accurate, particularly in the years furthest out; companies such as ExxonMobil made adjustments along the way. But the forecasts were necessary given the multibillion-dollar cost of oil and gas projects, in which even successes take decades to pay off. They helped the companies form a general picture of the world to come so that they could make coherent investments.

The outlook made notable prophecies for batteries and electric cars. It started by forecasting that oil and gas would supply 60 percent of the world's energy a quarter century ahead. That actually represented an increase from 55 percent in 2010. The prediction was almost categorical—the company foresaw no specific threat to this continued dominance. Biofuels, solar, wind, and other non–fossil fuels and technologies all seemed destined to remain permanently marginal.

But batteries were a different matter. Perhaps recalling the company's hasty surrender of lithium-ion to Japan three decades earlier, the ExxonMobil scenarists flagged batteries as one of the few wild cards with the potential to disrupt its world. ExxonMobil did not spell it out, but if researchers somewhere made a serious advance in battery technology, a jump in performance by a factor of four or five, they could grievously undermine oil. The car and truck fleet would require much less gasoline as consumers made economically driven choices to buy quiet electrics. The number of cars on the road around the world would still double to about 1.6 billion by 2040, but if many were electric, oil companies would have to become different animals, both smaller and sleeker.

In fact, ExxonMobil did forecast a shift to electrics of a sort—it thought that almost half the global fleet would be electrified in 2040. But most of these would be hybrids—glorified electrics like the Prius, with baby-size batteries that could propel a vehicle five or six miles with the engine shut off. Plug-in hybrids and pure electrics would capture 2.5 percent of the market. That added up to around 40 million of them, a highly impressive number. But it paled next to the *680 million* Prius-like hybrids that would be on the road. This was not a genuine electric picture.

It *was* a shift: at least two thirds of all cars sold after 2025 or 2030 would be equipped with some form of electric technology. And you could reach your own conclusions as to why it would take place. One significant influence would probably be consumer taste. At some point, Americans would probably reject pure gasoline-fueled cars, just as they largely stopped throwing garbage from their car windows amid Lady Bird Johnson's "Keep America Beautiful" campaign in the mid-1960s. Only a narrowing niche would even contemplate a model unequipped with some form of gasoline-saving electric propulsion. Other factors would contribute, too, but the main idea was that motorists would make the shift.

The trickier question was why, three decades in the future, the market for *pure electrics* would remain stunted. The answer, according to ExxonMobil, was that by numerous metrics important to buyers, electrics would barely close the gap with the internal combustion engine. In the current market, with gasoline at $3.50 to $4.00 a gallon, a Prius owner required just four years of fuel savings to cover the vehicle's higher price tag. Not so the pure-electric Nissan Leaf. It cost roughly $12,000 more than comparable gasoline-fueled models. Gasoline would have to cost $10 a gallon to compensate for such a price difference over the space of a five-year loan. So an electric-car owner would save on fuel but might never recoup the elevated cost of the vehicle. ExxonMobil forecast the persistence of this price difference more or less all the way through to 2040.

Wall Street aligned with the pessimists. About this time, Edward Morse, an analyst with Citigroup, forecast a fresh, decades-long period of relatively cheap oil ranging from $70 to $90 a barrel. He said this new day would be triggered by a surge of oil production and a moderation of demand.[1] If Morse was correct, American public support for electric subsidies could evaporate. China faced the same conundrum—already Wan Gang was threatening to slash China's subsidies.

ExxonMobil's logic was perhaps generally accurate—oil and gasoline propulsion were dug in. But could one be certain that time would stand effectively still in battery labs for almost *three decades*? That for the many years ahead, Thackeray, Amine, the many other scientists across the globe, and the generation of researchers to come all would fail to crack the battery code? The world was in a state of utter flux, of financial collapse, extreme weather patterns, and the fall of Middle East governments, not to mention a wholly unexpected—and big—shakeup from the shale gas and oil boom. How could one be certain of anything? The big necessary jump in performance and cost was formidable, but not impossible. Were it achieved, one of the greatest victims would be fossil fuels. Big Oil as we currently knew it would shrivel.

With respect to electric cars, ExxonMobil's forecast seemed built on presumptuous ground.

Menahem Anderman, an Israeli-born battery guy with a Ph.D. in physical chemistry from UC Berkeley, organized the Orlando conference. Anderman said he was "the world's leading independent expert on advanced automotive batteries." ExxonMobil's views about batteries were not altogether surprising, but Anderman's were—he, too, was skeptical of them, along with electric cars. Industry hands were paying him thousands of dollars each to hear that their mission was more or less hopeless, at least for a long time to come.

Anderman was distrustful of the entire thesis of the battery race. "Somehow," he said, "there was a decision during the 2009 collapse of the financial market and the auto industry that the solution was to electrify the fleet. But there was no connection whatever between the financial crisis, the automotive industry crisis, and electrification." He said the global pursuit was a mere gimmick to "create positive motivation with employees, suppliers, shareholders, with the public, with the press, and government." For three decades, electric-car proponents had awaited a convergence of events that would catapult the vehicles into the commercial market. But it simply wasn't to be—not this time anyway. Researchers might achieve a genuine breakthrough in a decade or so, Anderman said. But meanwhile the internal combustion engine would keep improving and "raising the bar."

30

The Old Technology Guys

I f you walked downhill for fifteen minutes from the Battery Lab, past some woods and to the right, you'd be likely to find Don Hillebrand in Building 362, one of a cluster of three-story steel-and-glass structures. Relaxed and plain-spoken, he grew up, attended college, and reached the age of thirty-one before leaving the rural Michigan county of Oakland to take a series of car industry jobs in Washington, D.C., and Germany.

Hillebrand frequently irritated the battery guys. His job was to oversee Argonne's work on advanced internal combustion engines. Hillebrand said the conventional engine had a lot of fight left "and will ultimately beat electrics for another generation." Batteries and internal combustion, he said, were on separate innovation tracks. When scientists like Thackeray and Amine made more powerful batteries, the internal combustion engine grew more efficient through tinkering. To close the gap, the battery guys had to devise better technology at a faster pace than internal combustion. So far the gasoline engine had managed to maintain its lead. That was what Anderman meant by combustion's capacity for "raising the bar." The rivals in the battery and electric car race were racing each other, and all as a group were battling agile internal combustion.

Recently, Hillebrand discussed the future with a group of overweight, grizzled men with lines under their eyes. "Kind of legendary" in the global car business, they were the vice presidents of major industry players like GM, Ford, Bosch, and Nissan, the men who, one step down from the CEO, decided what cars their companies actually produced. They tended not to "put up with any crap," Hillebrand said. "They are not interested in what sounds interesting and what sounds cool," he said, but in "things that are really going to happen." It became evident that they did not foresee a breakout of the electric car for many years to come. Electrics cost too much to produce. There was no indication that the economics were going to significantly improve. Motorists might keep buying 20,000 or 30,000 Leafs and Volts a year, they said, but there was no sign that either model would achieve the hundreds-of-thousands-of-cars-a-year sales that signaled mass appeal.

The old guys were right, Hillebrand said. He himself foresaw internal combustion vehicles that could run automatically on almost any fossil fuel. As it stood, mass-market diesel engines, relying on compression rather than spark plugs to ignite the fuel that drove the car, were probably the most efficient on the planet—fully 45 percent of the diesel poured into the tank ended up in the propulsion of the vehicle; just 55 percent burned off as wasted heat in the process of combustion. As for gasoline, just 18 percent of its energy actually reached the wheels; a whopping 82 percent went into the ether. That made diesels the vehicle of choice in Europe, with its higher-priced fuel, while Americans still by far favored the generally quieter and more powerful gasoline engine. But Hillebrand foresaw Americans catching up in efficiency because researchers would transform gasoline into a fuel competitive with diesel. If that could be done, and his own team at Argonne was working to make it so, the gasoline-fueled internal combustion engine would attain 35 or even 40 percent better mileage. If motorists maintained their current driving habits, that meant the world might burn 20 million fewer barrels of oil a day, more than a fifth of global consumption. The resulting global surplus would send down oil and gasoline prices everywhere. OPEC and Russia would suffer a severe blow to their income. Oil-short nations would benefit from sharply lower import costs. As for batteries, the bar would be even higher. "Taking the infrastructure we have, using engines we understand with no new costs, that is where we are going in the next fifteen years, and what is gonna compete really effectively with electrics," Hillebrand said.

Carmakers would still have to *produce* electrics—they would be necessary to meet American requirements for a fifty-four-mile-per-gallon average across the vehicle fleet. Highly efficient engines with direct fuel injection, slim, titanium bodies, and decreased power would get you a long way, as would hybrids and plug-ins. But you would still require pure electrics burning no fuel at all to reach the compulsory fifty-four. A fraction of the fleet, but still some electrics.

Such electric bearishness, incidentally, was not shared by the Asian car bosses with whom Hillebrand was speaking that day. By way of an informal, hands-in-the-air survey, the Japanese and Chinese vice presidents signaled a bright future for electric cars. They clearly saw themselves winning by dint of long, iterative patience and rejected American and German pessimism. They didn't share the same memories of the early twentieth century, when electrics died for the lack of adequate batteries and infrastructure, or of the 1990s, when the same happened to GM's EV1, the carmaker's abortive stab at electrics in the 1990s.

Hillebrand said that the question of who was right—the Asians or the Westerners—boiled down to whether you thought that this time was different. Paying six to seven dollars a gallon for diesel, as Europeans currently did, had not impelled them to shift to electrics, for example; instead, the Europeans were merely buying smaller diesel-fueled cars. Perhaps electrics would gain more favor if gasoline or diesel rose to ten dollars a gallon. But Hillebrand's main point was that the American fleet was "not going to electrify just because we want it to." It was a swipe at battery race enthusiasm, which again aggravated Chamberlain's guys.

This annoyance did not constrain Hillebrand. He suggested that the battery guys did not fully appreciate what was required to win. They and carmakers were rushing the science. Automakers were enamored with "bling," thinking they would earn street cred and general adoration with flashy but ultimately hollow products. The truth, he said, was that big advances took time. All but the most exceptional technologies reached the broad market only after slow, evolutionary steps. In the case of electrics, he said, the automakers would do best by optimizing small, battery-enhanced hybrids like the Prius until they matured into a standard vehicle feature. That would lead to better plug-in hybrids like the Volt. The plug-ins would gain widespread adoption. Then—and only then, perhaps two decades in the future—electrics might be ready for the broad market.

The battery race had to be won methodically. Moving too fast would place substandard vehicles on the road and give the technology a black eye. "Things you do to accelerate do not always accelerate," Hillebrand said. "Sometimes they give people a bad taste of the technology and then you have to go away for a while and wait until those people forget about it, and bring it out again, because it was not ready."

The shakeout was all around Argonne. EnerDel, an Indiana company that four years before shared an *R&D* Magazine 100 award with Khalil Amine, filed for bankruptcy and was on the brink of acquisition by a Russian timber magnate. A123—the darling of investors in 2009—furloughed a third of its employees as sales failed to materialize. People began to forecast its demise. Amine said, "A Chinese or Japanese company will buy A123 for nothing. Mark my words." He foresaw bankruptcy for a lot of start-ups and survival only for large companies with deep pockets and an independent income stream. Kevin Gallagher said, "I feel we are at the top of the bubble where it pops."

Such talk typified the usual path of new technology, Hillebrand said. To demonstrate, he drew a chart. It consisted of a bicycle riding over two hills. The ascent up the first equated to a concept's early, optimistic days. Near the crest, lavish funding arrived to push the idea along (this was the location of advanced batteries at the moment, he said). Then came inevitable setbacks, given that no early technology was perfect. This induced a downhill plunge, including loss of investor and public confidence, and a shakeout (such as electric cars were just then experiencing). The descent was long and miserable. But it inevitably bottomed out. If the technology was solid, you began to ascend the second hill. Along the way, problems were solved, confidence was restored. Ultimately, you conquered the market.

Consider the cell phone, electric car enthusiasts said. Nearly every teenager and adult in the world seemed to own a mobile device, with six billion in use. But the first consumer cell phone was introduced in 1983. It took another two decades for mobile phones to become commonplace, to reach the point at which they were owned by a majority of Americans. Some futurists said the technology cycle was accelerating, yet it was the same story with the smart phone; it took *nineteen years* to move from first product in 1994 (the IBM Simon Personal Communicator) to 50 percent penetration in 2013. By that measure, electrified vehicles had a ways to go—the first Prius was sold in the United States in 2001 and the Volt and the Leaf a decade after that. The doubters should probably lay off until well into the 2020s.

Amine said the despair around him overlooked how advances truly happened. "Scientists get stubborn," he said. "You have a hard goal that no one thinks can be done. But then some clever guy comes from nowhere and, bang, it is solved. Scientists have to be optimists."

Unlike microchips, batteries don't adhere to a principle akin to Moore's law, the rule of thumb that the number of switches on a chip—semiconductor efficiency—doubles about every eighteen months. Batteries were comparatively slow to advance. But that did not make electronics superior to electric cars.

Consumer electronics typically wear out and require replacement every two or three years. They lock up, go on the fritz, and generally degrade. They are fragile when jostled or dropped and are often cheaper to replace than repair. If battery manufacturers and carmakers produced such mediocrity, they could be run out of business, sued for billions and perhaps even go to prison if anything catastrophic occurred. Automobiles have to last at least a decade and start every time. Their performance had to remain roughly the same throughout. They had to be safe while moving— or crashing—at high speed.

Smart phones, iPods, and the like were elite and disposable fashion statements and not on the same technological plane as the electrochemistry underpinning advanced batteries. You installed smart phones and iPods *in* an electric car. They were mere devices, inferior accessories compared with the science and engineering underlying the lithium-ion battery.

Competitive consumer electronics played to the strengths of the Asian strategy of inching ahead through methodical improvement in engineering and manufacturing—the newest iteration of the iPhone sold wildly as a symbol of style. But the sluggish battery race might turn out otherwise. Even though the United States appeared to be near the back of the pack at the moment, its profound skills in the lab meant it could leapfrog over all.

Here was where Hillebrand collaborated with the battery guys. The Asian manufacturing giants—Japan and China in particular— stood accused of gaining their original edge by stealing Western technology. But all the leading nations reverse-engineered rival brands. In the United States, Hillebrand supervised such work and turned over the data to American carmakers. He called it a patriotic duty. Hillebrand favored a restoration of American manufacturing. Obama had identified advanced batteries and the electric car as the priority. Hillebrand was all in, despite his reservations about the chances for success.

Hillebrand strolled through an airport-hangar-size warehouse. The structure was among several from Argonne's first decade. When Hillebrand first happened upon them a few years prior, they had been crammed with magnets and nuclear equipment. He had four of them emptied and retrofitted into automobile labs.

"This is a Korean hybrid vehicle," Hillebrand said. He gestured toward a half-dismantled sedan. It looked a lot like the Prius. "They are trying to mimic the system," he said. Vehicles such as this were actually provided by the South Korean manufacturer. It understood that the data would be handed over to its American rivals, but Hillebrand would give a copy to the South Koreans as well, which was valuable information. ".They benefit, we benefit," Hillebrand said. "It is the proper definition of partnership."

Hillebrand reached an invention that he believed would confound electrics. It was called the "omnivorous engine." It was an old GM motor retooled to burn any type of carbon-based or synthetic fuel—methanol, butanol, ethanol, and of course gasoline or diesel. The engine self-adjusted for anything you put into the tank. "It is really a sleeper technology," he said. "It is going to have a huge impact."

He moved on. "Here is a Volt—we just got it," he said. "It is the most advanced vehicle in the world, the most fuel-efficient and possibly the best." He appreciated the Volt in spite of his confidence in the economic case for fossil-fuel engines. His guys were preparing to disassemble it down to its individual instruments, put it back together, test it, and take it apart again before putting it back together and retesting it. Then they would park it in a corner in case other questions arose. When they were finished, they would haul it to the dump. "Because by then, it is not a car anymore," he said. As of now, it seemed that GM had packed far more capability into the Volt than it was informing anyone. "When you buy that car, you are getting twice the value that you are actually paying for," Hillebrand said.

"This is the Prius," he said. The car he pointed to was a shell with wires hanging out. "It is an example of what they look like when we are done," he said. This particular examination had proven exceedingly useful because when the second-generation Prius was released in the mid-2000s, some wondered whether

Toyota had cheated on the fuel economy tests. Hillebrand's team had showed that, if the company wanted to, it in fact could game federal evaluators. That was because the car could be programmed with advance knowledge of the curves, stops, and hazards that all automakers knew the test featured. So armed, it could adjust and conserve gasoline. Hillebrand's team did not demonstrate that the Prius folks did cheat. But the opportunity to do so was sufficient. He sent word to the Environmental Protection Agency, which devised a randomized test that was harder to con.

The exercise of disassembling a selection of all the major vehicles on the planet taught Hillebrand that almost every new automotive technology followed a long adoption curve, including disc brakes, fuel injection, and the automatic transmission. He could think of no big advance that achieved immediate acceptance. For electric cars, the largest unknown was outside the control of the inventors or manufacturers. It was oil prices. If they were comparatively moderate for a sustained period, electrics would be even more hobbled than they already were. But if prices climbed and stayed high, creating buying anxiety at the gasoline pump despite improved internal combustion efficiency, they could motivate more concentrated research on better batteries and success in a decade. "We could do it if we have to," he said. "The problem has been that we haven't had to."

31

Only the Irrational or the Naïve Will Win the Day

I f the odds of beating gasoline were so low, how could the battery guys hope to win? To find out, Sujeet Kumar, Jeff Chamberlain, and dozens of private energy executives filed into a darkened conference room at UC Berkeley, invited by the Department of Energy for a two-day gathering just before Orlando. The morning keynote speaker was Vinod Khosla, the most aggressive clean-energy investor in Silicon Valley. The New Delhi–born Khosla had graduated from the elite Indian Institutes of Technology and gone on to cofound Sun Microsystems. In recent years, he had invested more than $1 billion of his own and investors' money in solar power, biofuels, and batteries. He was blunt-spoken and usually dressed in black.

Khosla began by challenging the premise of gasoline's invincibility: recent history showed that it was as vulnerable as any other incumbent technology, he said. That the odds of beating gasoline were low was precisely why he was invested in doing so—it made the potential financial gain astronomical. "Experts as a group speak knowingly of the 2008 financial crisis, but in June 2008 none predicted it," he said. The same was true for energy. Experts spoke with great clarity as to the future of shale gas. In 2008, none forecast its arrival.

"You say that I have a hundredth of a percent shot at success?" he said. "I'll take the odds."

A slide appeared on the wall. "For those who can't read it, the

probability of success is on the vertical axis and the chance of de-
structive impact on the horizontal," he said. "What I am saying is
that when there is more than a 90 percent chance of a technology
failing, that is when you tend to have the most disruptive potential." For venture capitalists, the destruction of an established pattern of business—and hence the creation of a new way—tended to
bring by far the largest profit.

That was not how most venture and angel investors worked, he
said. Instead, venture capitalists sought to reduce risk to a point at
which the difference between the consequences of failure and of
success was incremental.

"I am suggesting the exact opposite," Khosla said. One should
welcome acute risk and the potential upside the risk offered.

Khosla knew that his message, while heard by most or all of
those sitting before him, would be heeded by very few because
"only unreasonable and naïve people can attempt things that are
near impossible." Insults might at least discomfit them. "Experts
who can always tell you why something won't work—they can always scare reasonable, rational people from attempting these crazy
ideas," he said.

ExxonMobil did not entirely rule out Khosla's scenario. That
was clear on page 48 of its 2040 outlook.

"Technology also can be unpredictable," ExxonMobil's futurists
said. "A breakthrough in low-cost, large-scale storage of electricity
would greatly improve the prospect for wind and solar for electricity generation. Faster-than-expected drops in battery costs would
likely make electric cars more of a factor through 2040 than we
expect them to be."

In other words, the Argonne group—and any of the teams in
the battery race—could confound ExxonMobil's prognosis. The
outlook was where the company and Khosla converged. His core
investment principle was the oil giant's nightmare scenario.

32

A Three-Hundred-Mile Battery

Not everyone in Orlando was dejected. Kumar, for one, was not. Neither were the Japanese present at the conference. They believed the battery industry was not so much doomed as situated in the much-dreaded "valley of death." This was the gaping and long chasm between the completion of a product and its arrival in the marketplace. How gaping and long was unpredictable. The space of time could be months, decades, or unbridgeable. Start-up companies routinely failed during this stage, often for lack of cash or conviction. They lost to the market. It was why Amine said that many—perhaps most—of the electric pioneers were destined to this fate.

But the valley had another meaning as well, according to a number of voices at the conference, which was that those willing and able to hang on three, five, or nine more years could find themselves in a very different market.

One pathway to a seriously powerful battery was to improve the anode rather than the cathode. The anode was the staging point for the lithium. From there, the lithium shuttled to the cathode, providing the current that propelled an electric. Anodes were judged by how much lithium they could store and the rate at which it could be extracted, which was what delivered distance and acceleration. The standard was a graphite anode developed by Bell Labs back in the 1970s. Kumar was engaged in an industry-wide competition to replace it with an anode made of silicon, a metal that could absorb a much larger ratio of lithium atoms. A graphite anode absorbed one lithium atom for every six carbon atoms; but each silicon atom could accommodate four lithium atoms. Next to pure lithium metal, that made silicon the most energetic possible anode. Such an anode had the potential to deliver an order of magnitude better performance than graphite, whose discharge capacity was about 400 milliampere-hours per gram. Silicon had a theoretical capacity of *4,000* milliampere-hours per gram. You could not hope to attain that maximal peak in practice, but 1,400 or 1,600 were conceivable and, if achieved, would more than triple the graphite anode's performance.

But silicon had a problem. For use in an automobile, you needed an anode to withstand at least 1,000 charge-discharge cycles. As you intercalated lithium into a silicon anode, it expanded tremendously. Graphite more or less maintained its shape while absorbing the lithium, but silicon blew up three or four times in size. As you charged and discharged again and again, the anode kept expanding and contracting, until it finally pulverized and killed the battery.

This was not news. The virtues of silicon had long been discussed, but no one had yet managed to resolve the expansion issue. Researchers would reach two or three dozen charge-and-discharge cycles, and the anode would break up. Everyone knew that. But Kumar wanted to try his hand. The motivation was a competition run by ARPA-E, the Department of Energy's new funding unit for radical technologies. The grants, for ideas promising profound leaps in energy technology, ranged from $500,000 to $10 million. Considering that Envia's entire first round of funding was just over $3 million, the sum in play was beguiling. The prestige of an ARPA-E grant could also attract yet more money and industry attention to Envia's.

Kumar's best chance seemed to be twinning a silicon anode with NMC 2.0. Khalil Amine had an interesting concept for silicon, so Kumar applied for the ARPA-E competition with the Argonne scientist as a partner. The submission was straightforward. It named the Amine silicon concept plus a few others that, if coupled with the

NMC 2.0, could result in a 400-watt-hour-per-kilogram battery, sufficient to enable a three-hundred-mile car. That seemed to Kumar to meet ARPA-E's requirement for a transformational breakthrough.

It was a bold proposal—the generally accepted physical limit of a lithium-ion battery using a graphite anode was 280 watt-hours per kilogram. No one had ever created a 400-watt-hour-per-kilogram battery. In all, ARPA-E received some 3,700 submissions for $150 million in awards. Thirty-seven were selected. Envia was among them—Kumar won a $4 million grant.

For the subsequent year, Kumar's team worked through the handful of silicon anode concepts he had proposed until it settled on one. Kumar said Amine's anode, a composite of silicon and graphene, pure carbon material the thickness of an atom, had failed to meet the necessary metrics. Instead, the best anode was made of silicon monoxide particles embedded into carbon. Kumar's team built pores into this silicon-carbon combination measuring between 50 nanometers and 5 microns in diameter, and filled them with electrolyte. Carbon in the shape of fibers or nano-size tubes were also mixed into the anode, thus creating an electrically conductive network. The silicon's expansion was thus redirected and absorbed. Even if the silicon broke apart immediately, the carbon fibers and tubes provided a path across which the lithium ions could pass on their way to and from the cathode.

Kumar said the results were excellent but that there was a disadvantage to working at nanoscale. This path to the better battery was expensive. You started with a vacuum reactor and a costly substrate, sometimes using platinum, a precious metal. Then you grew nanowires and nanotubes. What resulted was like pixie dust—you derived just milligrams of material each time while what required was bulk powder. The process might decline in cost over time, but for now it could not be justified.

Perhaps there was a cheaper way. What if his team skipped the vacuum reactor and the platinum and instead employed a conventional furnace to transform a precursor of cheap silica into good-quality, ten-gram lots of powder, just enough to make coin cell samples? That would significantly reduce the cost. Kumar's team would try.

The jerry-rigging worked. Together, the cheaper components—including Kumar's version of NMC 2.0 on the cathode side—were delivering the milestone energy density of 400 watt-hours per kilogram.

When he heard of the success, Arun Majumdar, the director of ARPA-E, already a fan of Envia's, was elated. He said that Kumar should now seek independent verification. This was a very big deal and Kumar would want to validate his claim through an unimpeachable expert or body.

Kumar sent the material to Crane, a respected Indiana-based evaluation division of the Naval Surface Warfare Center. Crane came back with a stamp of approval—it had put the cell through twenty-two charge-discharge cycles and confirmed the 400-watt-hour-per-kilogram energy density. Kumar had reached the milestone.

It was a considerable achievement, perhaps big enough, Majumdar said, to justify its announcement at the annual ARPA-E Summit in Washington, which was just two months away. Former president Bill Clinton and Microsoft cofounder Bill Gates would both be keynote speakers. Majumdar said he would let Kumar know his decision.

The prospect of such attention animated Kumar and his partners. The spotlight would be of incalculable promotional value considering their aspirations for an IPO. Awaiting Majumdar's decision, Kumar traveled for a dress rehearsal at the Orlando conference.

Kumar walked on stage at ChampionsGate wearing a dark blue suit. Flipping through a slide deck, he said that what he was describing did not involve a typical laboratory experiment—it was not grams of material encased in a nickel-size coin cell, but a standard 45-amp-hour electric-car battery, vacuum-sealed inside a manila-envelope-size pouch. It was a breakthrough that could finally enable the long-range, affordable electric car.

Kumar recounted some facts about NMC 2.0—if you pushed the charging voltage to 4.5 volts, the blockbuster cathode, twinned with a graphite anode, delivered 280 watt-hours per kilogram, double the performance of the lithium-cobalt-oxide material contained in the audience's smart phones and laptops. But you achieved a wallop when you swapped in the silicon-carbon anode— 400 watt-hours per kilogram, which was "a world record," Kumar said. In addition, the advance reduced the cost to just $250 per kilowatt-hour, half that of lithium-cobalt-oxide. Just two years in the future, he said, the cost would be down to $180 a kilowatt-hour.

The battery was only a prototype—he had charged and discharged it just three hundred times. Experts in the audience knew that Kumar would have to more than triple the number of cycles before the battery could be used in a car. Kumar himself would tell you that climb would be "very tough. Very complicated." Another $4 million or $5 million in R&D spending might be required to reach all the way there. Challenges remained. But Envia was "breaking the barrier," he said.

One by one after the presentation, the representatives of GM, Dow Chemical, and Hyundai approached Kumar. They wanted a private word.

Anderman, the conference organizer, said Kumar was exaggerating. His talk had served a purpose, he said—it had demonstrated to the shell-shocked industry that companies were still trying. He had invited Kumar to speak for precisely that reason. "But there is no breakthrough at Envia," he said. Like everyone in the industry, Kumar was working to stop silicon from swelling, but he had not done so as yet. Neither had he achieved 400 watt-hours per kilogram. He was hyping his achievements.

Jeff Dahn, the Canadian researcher at Dalhousie University in Halifax, felt differently. One of the most original minds in batteries, Dahn was notorious for ripping into the ideas of his colleagues— publicly and usually with precision. He pointed out flaws that most battery guys, knowing how hard it was to make an advance of any type, typically kept to themselves. Dahn was with Anderman in the belief that battery scientists often cherry-picked their results in order to postulate nonexistent advances. That did not make him a pessimist—he was a true believer, confident that scientists would eventually get it right. To reach that point, they needed first to stop doctoring the results and be honest with the world and themselves. Dahn had recently delivered blunt PowerPoint presentations that, spliced with videos of exploding batteries, accused fellow scientists, including Khalil Amine, of camouflaging the risk that their inventions could catch fire. But Kumar, he said, was not a member of this group of embellishers.

Dahn was unusually complimentary to Kumar. What the Envia man unveiled was not necessarily elegant—it was really an engineering feat, packed together efficiently. But it also worked. "It looks like it can get to four hundred," he said. "I am very familiar with the materials that he is talking about and I think it is doable."

Dahn was acting coy. Back in 1999, the 3M Company had filed a patent application for his version of NMC just a few months after Argonne. In the subsequent years, Thackeray and Dahn bickered over the precise atomic structure of NMC—was it a saucy amalgam of nickel, cobalt, manganese, and lithium (Dahn's position), known as "solid solution," or a more structured composite with a discernible chemical architecture (Thackeray's)? For the motorist, the difference seemed to be immaterial. But it could prove crucial should NMC 2.0 become part of a massively best-selling battery. Thackeray had managed to win the crucial original American patent while the 3M Company had grabbed patent rights in China, Japan, and South Korea. 3M had gone to war with pilferers of NMC—Sony, Matsushita, and Panasonic—and won. The details of the settlements were sealed, but the outcome upheld the Dahn patents. 3M had filed no suit against Argonne, but Dahn made it sound like one was possible. He said, "I think Argonne and 3M are not on the best terms." It was only a matter of time before 3M went after GM, too. In Dahn's view, it was his IP and not Argonne's that was contained in the Volt. "I think that

Argonne is just using this composite argument to stay outside the 3M patent," he said.

When Dahn was younger, he could get worked up over who was tromping on his perceived turf, but he said he welcomed Kumar's work on NMC. There was an enormous gulf between achieving three hundred and one thousand charge-discharge cycles. "That is going to take some time," Dahn said. But for starters, three hundred cycles "look pretty darn good" if one was aiming at smart phones and laptops. Such performance would give longer life at the same cost as current batteries. "I am sure Apple would love to make the iPhone lighter and thinner," he said.

Was his position on Envia a more positive, new Jeff Dahn?

"No. I am being realistic," he said.

Do the math, Dahn said. The basic NMC-spinel battery in the GM Volt delivered about 100 watt-hours per kilogram. Since GM overengineered the battery to maintain a margin for error, about 37 percent of it went unused—the excess was there just in case added capacity was needed. So it was effectively running at about 66 watt-hours per kilogram. If you now doubled the capacity using the Envia formulation and slimmed down the unused capacity, you would triple your range—rather than 40 miles, the Volt would travel more than 120 miles on a single charge.

Alternatively, GM could stay with the 40-mile range and cut about $10,000 off the price of the car. "You have your choice," Dahn said. "This is why people are fighting for higher energy and longer life. It is what it is all about."

Dahn had questions. For example, why didn't Kumar report more data? What happened after three hundred cycles? But he was not worried—they were familiar questions. The first lithium-ion battery—Sony's 1991 technology—was "a piece of junk." But since then, its performance had improved by a factor of two, making lithium-ion tower over anything that existed previously. Envia's three hundred cycles would increase. "How long and how fast? Nobody knows," Dahn said. "But you can bet your bottom dollar it is going to get better."

33

ARPA-E

Two weeks after Orlando, Arun Majumdar presided over the ARPA-E Summit. He staged an atypical show for the unexciting Department of Energy. It opened with an appeal from Bill Clinton to Congress to significantly increase ARPA-E funding. The agency had much left to do and monumental gains to achieve, Clinton said. His image flashing onto three large screens within a large, darkened room, Clinton argued for ambition. Do not become mired in fretting. If you entered the room pessimistic, leave thinking of the grand possibilities, he said.

Clinton vanished behind a curtain and was replaced by Bill Gates and Steven Chu. They settled into armchairs. Gates was known for Windows, education, and health philanthropy, and not for his chops in energy. But Canadian energy thinker Vaclav Smil was his favorite writer, and Gates was a seed investor in a molten metal battery prototype invented by Donald Sadoway, a celebrity MIT chemist. Conversing with Chu, Gates said that clean power was perhaps the world's greatest challenge. It would be exceptionally harder than anything he himself had attempted. Gates said that when you contrasted energy and computer software, "people underestimate the difficulty getting the breakthroughs. And they underestimate how long it is going to take." Crossing from the invention to the marketplace was the longest wait of all—the general adoption of a new energy technology could take five to six decades, he said.

That's right, Chu replied. You could achieve an extraordinary advance, but no one could know when or whether it would be embraced by consumers. If you did not believe him, he said, go downstairs to the rows of new battery prototypes on display in the ARPA-E exhibition room. Single out the one that will capture the market and "go and invest in that." Such intuition stood little chance in deciphering the right electrochemical pathway.

Chu continued: "We need literally thousands of companies trying to increase the odds that we will end up with the ten or twenty approaches that will give us the magic solution." But the rewards for success would be astronomical. If someone for instance managed to combine breakthroughs in inexpensive batteries and solar cells, he said, the resulting invention would "go viral in the same way that cell phones went viral." It would do good in the world, since village populations in frontier countries could finally have power and electric light, and would create numerous vast fortunes. "Let's not blow it" through miserly energy research budgets, Chu said. "There is a huge market out there."

Majumdar took the podium. He claimed the keynote spot for himself.

Another alumnus of India's Institutes of Technology, Majumdar had previously been deputy director of the Lawrence Berkeley National Laboratory under Chu. In 2009, when Chamberlain held the very first pre-Hub meeting, Majumdar led the Berkeley delegation alongside electrochemistry pioneer John Newman. Although not a Bell veteran, a Nobel laureate, or a scientific impresario, Majumdar brought something more to the stage. Chu projected charm and charisma but Majumdar, resonant, grave, and confident, was theatrically masterful. In the basement, he said, were exhibitions of 180 select inventors—"the Wright brothers, Borlaugs, Salks, and Teslas of the twenty-first century. They are the crown jewels of our nation." They would help create the future. Nine merited special attention. He began to tick them off, flashing slides to punctuate their achievements.

About midway through, Majumdar said that a battery powerful

enough to propel a car the 225-mile distance from Washington, D.C., to Manhattan would cost $30,000. "Just the battery pack," he said. Since very few people could afford a car containing such a battery, ARPA-E had challenged the research community three years earlier to create a battery that could compete with gasoline in range and cost. Then, he said, "Let me tell you the story now of Envia."

A photograph of Kumar and the Envia team went up on the triple screens. The day before, Majumdar said, this start-up company had announced "the world record in energy density of a rechargeable lithium-ion battery." Its 400-watt-hour-per-kilogram battery, if scaled up, could take a car that entire Washington-to-New York journey in a single charge at half the cost of the current technology. And more was coming, he said.

The New York Times, granted an exclusive the previous evening, seemed uncertain how to treat the news. Its cautious headline read, "ENVIA CLAIMS 'BREAKTHROUGH' IN LITHIUM-ION BATTERY COST AND ENERGY DENSITY." Restraint vanished in the hours following Majumdar's presentation. One reporter declared Envia "the Golden Child" of the summit. Everyone was discussing the company, the story said. *Scientific American* recalled Envia's humble beginnings in the Palo Alto Library and calculated that its battery could fuel a three-hundred-mile car trip from St. Louis to Chicago for $10, an eighth of the cost of a gasoline fill-up for the same journey.

In the audience, the Argonne battery guys cringed. Then they went ballistic. Kevin Gallagher said Majumdar's claims about Envia were "bullshit," making him wonder about the other eight start-ups that he showcased. ARPA-E as a whole, with its pressures to deliver big leaps, was "basically set up for companies to lie," he said. Chamberlain didn't go that far but said that deceit was in the DNA of start-ups and VCs: you needed that quality in order to raise funding, sell your product, and ultimately achieve a successful exit—to flip your company in either an acquisition or an IPO. There was no blaming anyone for this Silicon Valley peculiarity.

They knew their remonstrations rang of sour grapes. For Argonne, not to mention others in the industry, the situation was confounding. How was Envia, a lab with three dozen researchers operating on a comparatively shoestring budget, managing advances that surpassed everyone else's, including the inventors of NMC 2.0? How had the company eclipsed the other NMC licensees, not to mention the Asian giants, all of whom were also working on both voltage fade and the silicon anode? Kapadia partly credited good office politics: All subjects were fair game—as long as they were respectful, any scientist could and often did debate any other. Anyone could attempt any experiment that he wished. Money was not held tight—they had $650,000 a month to spend and as long as the staff did not exceed it, no one challenged expenses. Kapadia said he worked "seamlessly" with Kumar, who was "a genius."

Did such operational conditions truly explain everything? Argonne's battery guys continued to grumble. They were not alone—Nissan and Samsung had come back to Envia with complaints that test cells provided by Kumar had swollen up with an apparent internal gassy buildup, and fresh samples had to be provided. In an unguarded moment, Envia cofounder Mike Sinkula had mentioned to Kapadia that their proprietary anode actually contained some Japanese material—it was not entirely Envia's invention, as the company's promotional material would easily lead one to believe. Kapadia was sufficiently concerned to run the claim by Purnesh Seegopaul, a materials scientist who served on the board. Seegopaul told him not to worry. "Great entrepreneurs bluff their way through. Look at Steve Jobs," Seegopaul said.[1] If Kumar was somewhat exaggerating, that was part of the game. Soon enough, Envia would invent an anode that was entirely its own. Plus, what was the probability that the skeptics were right? Kumar's peers were vetting Envia's material as part of federal grant rules. The world's biggest automotive companies were evaluating his cathode. Crane had weighed in. If Kumar was exaggerating, it seemed he would already have been found out.

Kapadia decided that Seegopaul was right: Envia continued to be ahead of everyone; such sniping was to be expected.

Chamberlain sank into a soft chair in the basement of the ARPA-E conference. The nearby cavernous exhibition hall was packed with the wares of ARPA-E winners, also-rans, and would-bes. He mulled over Envia's recent history and decided that Majumdar's high-profile announcement was politically driven. Department of Energy investments were a primary target of harsh Obama critics. The furor centered on Solyndra, a California solar power company that was awarded a $535 million stimulus loan and then filed for bankruptcy. Solyndra, critics said, exemplified the folly of "picking winners"—of favoring specific companies rather than general swaths of potential economic prosperity in which any enterprise might emerge a success. The loan, they said, was particularly suspect given that a Department of Energy official handling it was simultaneously a presidential campaign fund-raiser and married to a Solyndra lawyer.

In fact, ARPA-E and other programs *were* picking winners. But that was what they were supposed to do. The question was whether they picked wisely. In any case, while the wisdom of the Solyndra loan was debatable, its origins were in the Bush administration. ARPA-E was modeled after DARPA, a Pentagon research agency whose mission was to fund highly unusual ideas that otherwise might never be tried out. Solyndra, with its bet on a niche thin-film technology, fit that operating principle. The solar market had turned against everyone.

The politics remained unforgiving. In Washington, Republican congressmen held hearings accusing Solyndra executives and Obama officials of corruption. American voters would go to the polls in nine months. Obama could not be sanguine about reelection given these attacks and the listless economy. Against this backdrop, Envia was much-desired good news.

The announcement was extremely welcome at Envia. Kumar

and Kapadia suddenly had the rare attention of the world's media. As Chamberlain said, "It doesn't hurt to have a splash in *The New York Times*."

Chamberlain's main preoccupation was the Hub proposal, Steven Chu's initiative that would attempt to invent a new generation of batteries in collaboration with industrial partners who would manufacture them. The Department of Energy would select the site of the Battery Hub in a competition, and Chamberlain was working on the proposal in his hotel room. He said that about a hundred researchers and companies had expressed hopes of collaborating with Argonne. In addition to his core team, Chamberlain welcomed outside players because it was almost certain that three years or so down the road—if Argonne won—its plans would change, perhaps significantly.

This was why, in twenty minutes, Chamberlain and Gallagher planned to stroll the ARPA-E exposition and attempt to strike affiliate agreements with battery exhibitors. When advances were made, Chamberlain would possess a directory of who could make good use of them. "If our core partners aren't interested, we already have an audience that we can say, 'This is for you,'" he said.

Gallagher arrived and he and Chamberlain disappeared into the hall.

Kumar stood a few feet away, beaming and clutching a glass of wine. Men from rival companies and reporters crowded Envia's booth. Previously, Envia had been covered only by the Silicon Valley and energy blogs. Now *The New York Times* was calling.

"Hey, Atul!" Kumar shouted.

Kapadia walked up.

"It is crazy," the Envia CEO said. He had been fielding most of the media calls. His objective was to parlay the coverage into a commercial frenzy among carmakers over the Envia material. With the 300-mile range forecast, the calls weren't coming only from the media.

Majumdar had defined the race as a determination to keep advances such as Envia's in the United States. "If we don't act now, many of these innovations will go overseas and be manufactured elsewhere in the world," Majumdar said. Kapadia repeated that Envia in fact was in play. The Japanese automotive companies had initiated talks. After the Orlando conference, Kumar had also heard from eager managers at South Korea's Samsung and the American-Japanese joint venture Dow Kokam. Kapadia said, "If America does not commercialize this technology first, the question [that] should be asked in Detroit and D.C. is, 'Why?'"

General Motors was his and Kumar's first choice as a commercial partner. GM had invested first, winning it a two-year head start on everyone else. That was "thanks to Jon Lauckner's foresight," Kapadia said. Two days earlier, GM had awarded Lauckner a substantial promotion—he was now not only head of GM Ventures but also chief technology officer. Kapadia said that Lauckner needed to press his advantage if he was to win the industry rivalry to dominate electrics.

Lauckner was still in the lead. A week and a half earlier, a GM team had called on Envia. Kumar presented his latest voltage fade results along with fresh tests showing that the start-up's material met endurance criteria. Judging by the cells, the material could last fourteen years, longer than the ten years required under standard industry specs. A GM man said that an additional six-month test was required, but Kumar's sense was that licensing was only a matter of time.

Kapadia called Lauckner a "patriot partner." When the Envia CEO traveled to Japan, Lauckner peppered him with e-mails. His main message was not to sell the company to a Japanese or Korean buyer. "He will do whatever it takes with Envia to serve GM's purpose. And we like him for that," Kapadia said.

Envia had no intention of selling to a Japanese, Korean—or American—buyer. After months of consulting with Goldman Sachs and Morgan Stanley, Kapadia had decided that an IPO was the best option for Envia's investors. Envia would remain conservative—it would sell just 8 percent of the company on the market. The idea was to raise $60 million by selling two million shares at $30 each.

Kumar hoped to use the money to make a manufacturing plant. But the pair's goal wasn't just the plant: it was the valuation. Such a sale would factor out to a company value of $750 million. It wouldn't be quite the $1 billion of which he and Kumar had spoken, but it was a large sum. And perhaps, if a sufficient marketing effort was made, those shares might be sold at $40 each, which *would* achieve the aspired valuation.

Everything seemed possible. The decisive factor again was Envia's potential customer list—the four big carmakers from whom Kumar had received spec sheets. The Goldman and Morgan Stanley men had studied the prospect of licensing to these companies and said, "You must be kidding me. Don't even think about selling the company." Kapadia could and should take the riskier but potentially far more lucrative step of going public.

The 400-watt-hour-per-kilogram announcement was the first step. Now, Kapadia said, the senior managers of all the major car, parts, and chemical companies would be querying their staffs, "Are you working with Envia or not?" Kapadia said. Envia aimed to be to the 2010s what IBM was to the 1970s: the hottest stock on the market. By the end of the year, he expected to announce licensing deals with GM and Honda. They would move Envia into position to launch the IPO, which would take place "early '14, late '13," he said. Though Kapadia aimed for IBM's trendiness, he didn't aspire to its size. "We will stay small. Even as a public company I don't see us as more than one hundred people," he said.

Kumar and Kapadia beamed. "Success is a good feeling," Kapadia said.

34

The Old and the Young

One afternoon in the summer of 2012, Kevin Gallagher and his wife, Sabine, put out hummus, turkey sausages, and carrots in their backyard in Downers Grove, the suburb of choice for young Argonne researchers since the move from the University of Chicago in the 1940s. Lynn Trahey, a chemist with whom Gallagher shared an office, sipped a glass of home-brewed beer. Mike Slater, a postdoctoral chemist with a goatee and ponytail, juggled bowling pins. It was a few weeks after ARPA-E's summit and Gallagher was still irritated about Envia. He did not desire a public argument over the matter but said again that Kumar's 400-watt-hour-per-kilogram disclosure was just show. Gallagher was disposed to irritable pessimism—Thackeray said that was to be expected since he was an engineer. But he defended his suspicions on the basis of the girth of Kumar's electrodes: in order to deliver the performance that Envia claimed—meaning that an electric car could travel three hundred miles on a single charge—he would have had to densely pack the lithium into an unusually *thick* cathode. That was the only way. The problem was that thick electrodes were a blunt-force method—they could deliver the distance, but only in the lab. They probably could not be placed with confidence *into* a three-hundred-mile electric car. Being so fat, they would suffer early and fatal maladies and die long before the ten-year life span required for such batteries. They might even shatter. The future, Gallagher said, was slender electrodes—cathodes less than one hundred microns thick, or slimmer than the diameter of a human hair. In its rush to the market, Gallagher said, Envia had unveiled an attention-grabbing but flawed product that still required fundamental improvement.

Gallagher had grown up in a Detroit suburb called Trenton and by curious coincidence graduated not only from the same Georgia Tech doctoral program as Chamberlain, but also from the same elementary school. Both played the trombone and had a natural confidence. Chamberlain took a liking to Gallagher, who was fifteen years younger. He called him "the next generation." A lot of the other battery guys went along with that appraisal of Gallagher, who had curly brown hair, dimples, a compact, athletic build, and a smooth manner of speech. His first job out of college, before he decided to seek a Ph.D., was applying the adhesive to Scotch tape in a factory in Hutchinson, Minnesota, a town outside Minneapolis. He was the first to say that he could lose his temper. He had punched a man only once, and that wasn't really a man but his older brother Sean, whom he smacked in the face when they were boys. But when he did anger, it was when he felt one or more of his principles under attack. Chamberlain said that they were alike in this respect as well.

Perhaps Gallagher's outsized ire about Envia was another sign of the general malaise in batteries at the moment. Or possibly it was his idealism—he more than any of the battery guys was uncomfortable with the carnival aspect of batteries.

Lynn Trahey called Gallagher "K-Funk." She had joined Argonne three years earlier as a postdoc from Berkeley. Scientists in the United States were not only largely foreign born, but also mostly men. So Trahey was an anomaly on both accounts—she was the only female staff scientist in the Battery Department. She had been a cheerleader and played varsity doubles tennis in high school. As a graduate student, she wore a purple- and green-dyed ponytail. Trahey's current toned-down style appeared aimed at reducing her conspicuousness among these mostly plain men. She tied her hair back, unadorned. She dressed like one of the guys in loose-fitting jeans and sneakers. She seemed to overcompensate as well, in slangy and coarse phrases. "I avoid the small bathrooms near the Y-wing auditorium," she said once. "I think that's where old guys must go and take a long shit. I've seen rolled-up newspapers behind the pipe."

None of it worked. Trahey still stuck out. The guys behaved bizarrely around her. They spoke inexpressively, almost robotically. Except for Gallagher and Mike Slater, a lot of them simply stayed away. While colleagues behaved awkwardly, she was ideal for public relations exercises. At Berkeley, her professors dispatched her on community-outreach visits to neighborhood schools and senior-citizen groups. She would show up and attract favorable press for the department. Chamberlain employed Trahey to the same advantage. He featured a photograph of her posed in protective glasses on the department's home page and in a handful of press releases.

Another of the battery guys who conversed with Trahey was Dieter Gruen, the German-born physicist who had been at Argonne since the Manhattan Project days. Gruen was developing a concept for a lithium-sulfur battery and this summer asked Trahey's help with an experiment. Gruen had no funding for his idea; Trahey did, and, in an e-mail, he asked whether she would mind carrying out his experiment on the beam line.

Trahey was sitting in the office with Gallagher. She had already done one experiment for Gruen "to be nice," she said, but now "he is taking advantage." Gruen's work "is not in my budget."

"Why don't we get rid of the old people" at the lab? Gallagher said. "I'd like to see their output. I'll bet it's low." He said that if you calculated the average age of the department's researchers, you might be surprised as to how elderly the staff was as a whole. Gallagher and Trahey agreed that their older colleagues were costing too much money.

Trahey said, "The reason there are so few jobs is these people won't leave. These guys suck up all this money that could go to other things." It particularly galled her that Gruen was paid at the lab's top salary rank. "He is a 710!" she said.

Such grousing poured out of the pair. They suggested that

battery science was a young person's game. But were the ideas developed by over-the-hill scientists under scrutiny, or was it simply their ages? Judging by space and budget, it was true that the department could be insufferably cramped and the funding slender for fresh minds leaving the universities. But experience had its place, too. John Goodenough, the eighty-nine-year-old father of lithium-ion batteries, still ran his lab at the University of Texas. No one discussed pushing Goodenough into retirement.

Nor did Trahey and Gallagher speak that way of Goodenough's former protégé, Thackeray, who apart from also being a 710 was, at sixty-four, no longer a young man either. Sometimes Thackeray hinted at health problems but said it wasn't anything serious. That was understating it, as he had been treated for leukemia the previous year. But he said it was "not the aggressive form" and was more or less under control. "I'm not as well as I was ten years ago but I'm hanging in there," he said. Thackeray said he had no plans to retire soon and wondered how he would find time for a planned memoir of his South Africa days. As of now, he was occupied with voltage fade and administrative duties associated with managing two research groups. It could be some time before he got to the book.

Gallagher in fact owed his job to one of the old people. Paul Nelson had worked at Argonne since 1958. He was eighty-one years old, erect and fit. A few years before, a couple of the senior managers along with Dave Howell, the top battery boss at the Department of Energy, had decided that Nelson possessed a special skill that ought to be passed on—he had developed novel software that could rapidly assess whether a given new battery formulation would actually work in a vehicle. They searched awhile for an apt protégé before Gallagher's CV arrived in the mail. They hired him more or less on the spot.

Nelson had witnessed much of Argonne's battery history—the struggles with high-temperature sulfur batteries in the 1970s, the recruitment of Thackeray and Amine, and the advent of lithium-ion. By the early 1990s, he was nearing retirement age. At Argonne, that meant time to move on. A mere glance around told you that no one would appreciate his management advice. But Nelson did not want to move on. He was only in his early sixties and, as he saw it, had many productive years to go. He knew how to obtain an invitation to stay. If you proved your worth again as a pure scientist, you could be welcome back at the bench. Such arrangements were useful for Argonne because it tapped the product of a seasoned mind at a much smaller, part-time wage—really a symbolic sum. But you could stay in the game.

That's what Nelson did. He detected an opening in the assembly line that battery science had become: propose a new chemistry, obtain funding, prove (or fail to) that it worked in coin cells, write a paper, garner any accolades, then move on to the next thing. At no point was your idea typically tested for practicality—no one checked whether it could produce a superior battery. It was as though experimentation was the final product.

Nelson wanted to make it otherwise. Reality testing should assume a native role in the process. For the final three or so years of his tenure as Battery Department head, Nelson tailed off his supervisorial time and started to hone his skills at the computer. By the time he retired in 1995, he was adept at modeling and design, a new discipline in which you recreated an invention from scratch on the computer screen. Shifting to half-time, he ultimately developed the appraisal software, which, starting with a few grams of a material, could determine how it would perform in a complete automobile battery pack. He would calculate the weight and volume of all the ingredients—the lithium needed at the pack level, the electrolyte, steel, plastic, and so on—and extrapolate whether a lab-scale success would work for the decade or so a battery might be used out in the world. That was powerful, because no such system existed at the time. Nelson then expanded his work to also estimate the cost of manufacturing all these materials. He began to factor the price of the composition and the economies of scale—how much would be saved when you were producing, say, one hundred thousand batteries versus just a few hundred?

It was about this time when Gallagher was hired to learn Nelson's system. The pair began to meet with automakers and battery companies in order to obtain more precise estimates. The software's accuracy was soon impressive. When the pair decided to give it away on the Internet, posting it on the Argonne Web site, they attracted attention from industry players who perceived a new way to cull practical ideas from the less so. Other labs, including MIT, adopted it, too.

The Nelson-Gallagher model was far from standard procedure at Argonne itself. The senior managers understood its merit, as did their supervisors in Washington. But if you asked Amine whether he had vetted his new ideas with Nelson, he would stare as though you had suggested checking with Santa Claus. Trahey, Gallagher's own office mate, never inquired as to the feasibility of her concepts. That depressed Gallagher.

"The group is not interested in seeing how their work fits into the big picture," he said.

Trahey said that the other battery guys were simply "overwhelmed" with their various responsibilities. "They care. They don't want to be working on something futile. They *want* to know if it is something totally futile," she said.

"I don't think they do," Gallagher said.

Gallagher was right. Trahey was merely trying to salve his feelings. The model was sly. To the degree it was actually applied, the software could undermine the proclivity of some battery guys to inflate the potential of their inventions. It amounted to a stern manager peering over their shoulder.

As Gallagher understood it, the Hub would change all that. It was meant to be a dynamic research environment in which, if an idea was not working, resources would be more or less automatically redirected to something that might. To Gallagher, the Hub was a magical thought: Argonne *had* to win it. And he *had* to be in it. Here would be an honest appraisal of the battery work. The battery guys might never attain the desired electric-car performance; they might not win the battery race. But if they could not be straight with themselves, they *definitely* would not reach the goal, he said.

Gallagher's idea, embraced by Chamberlain, was to embed the vetting model into the Hub's management system. Armed with this data, a battery guy could peer deeper and isolate the specific materials having the best physical properties for the desired performance. The managers began to call the work "techno-economic modeling." Nelson's name dropped out of the discussion. Gallagher himself mentioned his former mentor's role almost only in passing. It wasn't that he was not grateful. "But at some point you have to hand it off so young folks can run with it," he said. Nelson did not seem to begrudge the treatment. It's what it meant to be one of the old people.

Gallagher was still uncertain about what the model could achieve. It was one matter to write such a system into the Hub proposal. It was another "to actually go and do it." It would be a new way of business for most of the lab and would require strong leadership to become an accepted tool.

One thing he knew was that most of the researchers did not believe that vetting was "as big of a deal as it actually is." They did not understand that if you could determine what did and did not work, that was "a clear stepping-stone" to the better battery. Perhaps the Hub would change their minds. The Hub's industry partners were certain to press the case. They would be biased toward such testing and intolerant of any inclination to "hide behind things. 'Cause if you make a bad product, everyone knows it," Gallagher said. "That's the thing—in the labs, you can hide behind things. You can play games."

Gallagher said that Envia's new battery should be subjected to the model. He had been in touch with Sun-Ho Kang, the former Argonne researcher now at Samsung. Kang's supervisors were inquiring about the ARPA-E announcement, and he wondered what Gallagher thought of the reported advance. Gallagher said he was skeptical. But he had managed to create philosophical distance from his early rejectionism. "It's natural that you say two things," he said. "First, 'Why couldn't I get those results?' Second, 'It's great that Envia did' because you had a role in creating this material that will be in the market." But one way to know for sure would be to examine the detailed data.

If the techno-economic model upset some people, "well, they shouldn't be in the Hub then, right?" Gallagher said. "The Hub is supposed to be a different way of doing business, and if you're not interested, then you don't have to be a part of it."

Gallagher's conviction was that Chamberlain and the other senior managers were not the only ones with careers on the line. Here was his chance to spend the next part of his life encouraging scrupulous work.

Chamberlain pulled up a story on his screen. That morning, gas had exploded in GM's experimental battery lab outside Detroit, blowing out three windows and forcing open a fortified door. Fortunately, no one was killed and just one worker had to be hospitalized. It happened during the extreme testing of an A123 battery under consideration for GM's future pure electric Chevy Spark.

"This isn't good," Chamberlain said.

The United States was not alone in its suffering. Notwithstanding the bravado in Orlando, Japan was experiencing angst over the demise of "Panel Bay," a densely packed strip of electronics factories three hundred miles southwest of Tokyo on Osaka Bay. After years of success, Panel Bay was now seeing Sharp, Panasonic, and other plants either shut down or sold off to Taiwan and other foreign buyers. Sony had lost its way. The edge was now held increasingly by South Korea and China.

But this did not cheer the battery guys, especially since the bad news was still piling up. Five days later, a forty-eight-page report by the Union of Concerned Scientists cast doubt on the green footprint of electric cars. The report noted that the cleanliness of a particular electric vehicle depended on where it was charged up. If its owner lived in a coal-burning state, the coal used to produce the electricity that charged up the car might be responsible for even greater emissions than a gasoline-fueled model. That is because oil and gasoline generally produce just half the CO_2 as coal. If the same electricity was created in a natural gas–burning plant, it would be a different story because it emits only one third of the greenhouse gases as coal. The report said 53 percent of Americans resided in coal-burning states. A *New York Times* headline asked, "HOW GREEN ARE ELECTRIC CARS? DEPENDS ON WHERE YOU PLUG IN."[1]

35

Red Team

The Hub was intended to emulate Bell Labs. But it also had to eclipse Bell's achievements. It had to both create advances *and* parlay them into commercial products. Chamberlain said it could not "do science only for the good and noble sake of science."

Eric Isaacs, the former Bell manager who ran Argonne, said the labs had their own Achilles heel. They failed to kill theories that weren't panning out, what industry players called "down-selecting." As the techno-economic model had highlighted, a battery guy typically conceived of a new electrode, attracted Department of Energy funding, and continued developing it as long as the money held out, whether or not it showed good progress. Consider voltage fade—it took the Battery Department more than a year to organize into a working team after Washington ordered it to. Chamberlain said, "That would never fly in industry."

The difference, Chamberlain said, was how government and industry scientists were evaluated: In the Battery Department, you were measured by the number of papers you published, how many times others referenced your work, your awards, and, in rare instances, the transition of your stuff into the market. In industry, the marker was singular—the company's financial performance.

GM's Mark Mathias worried that if the Department of Energy was not careful, the Hub could "end up being a bunch of research papers." Mathias led the automaker's electrochemical research lab outside Rochester, New York, and exercised considerable influence over whether GM would participate in the Hub. In order to succeed, he said, the Hub would have to be "run a little more like a business than is traditionally done in the national lab." Otherwise there was no point to collaborating with it.

There was, of course, the converse risk of down-selecting excessively, not providing the chance for promising ideas to bloom. The Hub had to settle somewhere in the middle—it had to cultivate creativity while ultimately culling out what was not working.

Chamberlain intended to "make swift decisions when we see something succeeding or we see something failing." That was not his current reputation. He was known to go to great lengths to get along. If he was deputy director of the Hub, which seemed likely, he would at once be thrust into the position of hatchet man—he would have to kill projects. But he said he was ready. "I'll do it," he said. Knowing himself, Chamberlain had organized the ground in advance so that he would not appear to be an ogre. He embedded incentives in the Hub structure to encourage down-selecting, or at least lessen the resistance: if your project failed—if you thought it should be killed—you could say so yourself and obtain first dibs on the next big high-risk project. The Hub would thus reward a researcher's proactive exit from an underperforming concept. Chamberlain would not have to be the one to say no. But it was a fair question as to how long he could maintain that front before he was found out and had to make clear his decisive hand in the down-selection.

Thackeray and his wife, Lisa, were sitting down at home for a dinner of pasta. His cancer had recurred, engorging his spleen with blood and distending his stomach. His doctor had prescribed chemotherapy and he had just undergone his final treatment. He spoke slowly and after dinner stretched out on a La-Z-Boy. Some 40 percent of his blood was concentrated just in the spleen, leaving the rest of his body to survive on the remaining 60 percent. "I feel guilty. I feel I have let everyone down with the Hub," he said.

Isaacs, the Argonne director, had wanted him to run the Hub, but
he simply wasn't up to another intensive five-year job. Isaacs seemed
to understand, but Thackeray still often raised the subject.

Regardless of the result of the chemo, Thackeray's doctor was
recommending that the spleen be removed. He wanted to conduct
a biopsy. If Thackeray had the type of cancer that the doctor suspected, the organ was better out. If he did not, there would have
been no need to remove it. But all in all, Thackeray said, "they say
it is better to have it out." One could live without a spleen. And the
doctor promised that he would "feel like I did five years ago." A
sparkle seemed to appear in his eye. His first plan after recovering
was to start on his book.

Isaacs was thinking ahead regarding the Hub competition. In
order to help ensure that Argonne did not blow it this time, he
built professional critics into the proposal writing process. These
were gadflies practiced at organizing militaristic "Pink Team" and
"Red Team" reviews and ripping proposals to shreds in order to get
them right.

Holly Coghill was an expert manager for big-ticket Department
of Energy proposals. She was steeped in the national labs—her
uncle once ran the Pacific Northwest National Laboratory and
Mark Peters, Isaacs's deputy, was a cousin. She knew the inside
personalities like they were family, including what they favored in
such proposals and how they preferred them to be organized. A
blonde with short-cropped feathery hair, Coghill favored Levi's
jackets and sent out a signal not to trifle with her. One Hub inter-
locutor, Coghill said, was "all sizzle and no stink. There's just not
a lot of depth there." When in her view an outsider spoke disre-
spectfully to Chamberlain, she replied, "Don't mess with my kids
or I'll freakin' kill ya."

But Isaacs's main secret weapon was Bill Madia, a nuclear
chemist who had run two national labs and established a victory
streak in the competition for large government projects. The sixty-
four-year-old Madia was an inflated, difficult-to-like egotist with
little apparent interest in getting along apart from with those

possessing influence over his next project. But he also was "the godfather of the national labs," Coghill said. He was ingenious at winning big-money, prestige-building competitions, making people forgive his foibles. This promised land of scientific shrine making was so coveted among the research elite that those able to lead the campaigns earned enormous fees and permission to exhibit almost any pathology. In all, the gadfly budget for Argonne's Hub proposal was about $500,000.

Isaacs was relying on Madia. "He has won multiple times," Isaacs said. "He just gets it." Isaacs wanted to win. He wanted to rid Argonne of its long, unlucky history. Madia was going to get him there.

He knew Madia's habit of making others miserable had already triggered an uprising on Chamberlain's team. They wanted him out. But Isaacs was unmoved. Madia was indispensable. If he was tough, Isaacs wanted him to be. "The most important thing," he said, "was (to make) sure that Jeff was on his toes."

The Pink Team's duties were clear. Its members were to appraise the draft proposal and provide recommendations on how to improve it, as a preliminary to a later, more exacting "Red Team." They also were to "score" it. Department of Energy judges evaluated proposals according to criteria set out in the Funding Opportunity Announcement, the contest declaration known as a FOA. The criteria—minutely detailed under broad categories such as scientific and technical merit, competency of the team, and reasonableness of the budget—were scored on a scale of one to one thousand. The Pink Team would provide Chamberlain with an assessment of Argonne's current standing.

The assessment lasted two days. Madia was harsh. Argonne's vision—its "story"—did not shine through. The narrative was "buried deep in the science." The scientific sections were adequate as far as they went, but the team's priority was to craft the story so that a nonexpert—like members of the judging team who were not battery guys—could understand it. As the proposal stood, it failed to meet this standard.

More problematically, the proposal seemed actually to ignore some provisions of the FOA. The FOA had stipulated a serious commitment to applied science. Madia judged the appropriate balance at about 60 percent research and the rest development and deployment. The Argonne team had proposed an 80 percent emphasis on basic research—clearly too much.

He raised a couple of other points—there were too many "whats" and not enough "hows"; each time the proposal said the team intended to do something, it should provide an example of how it would be done. Madia was troubled. The previous summer, he had seen a preliminary draft and said much the same. "You're no further than you were nine months ago," he now told the group.

Chamberlain's face went crimson. He was "mad as hell," Coghill said later. After a while, he collected himself. But Madia did not back down. There had to be "significant revisions," he said.

In its current shape, the proposal did not merit scoring. And he would not do it. The best course was to start rewriting.

Chamberlain e-mailed Isaacs: I think Madia is trying to manipulate us and we need to understand his motives, he wrote. Isaacs knew that Chamberlain's team was "sore as hell." He responded by spelling out explicitly what until then had gone unstated: Madia was an adviser. Chamberlain was in charge.

In the hours after the thrashing, Isaacs dined with Madia, who openly questioned Chamberlain's readiness for the project. The leader of such a proposal team was analogous to an air traffic controller, he said. He had to keep thirty-one planes in the air, ensure that they did not crash into one another, and hold them in their slots when one wished to land. Such navigation required alacrity. Unlike airplane pilots, who always followed instructions, Chamberlain had to preside over autonomous players. The battery guys alone included Stanford's Yi Cui, Berkeley's Venkat Srinivasan, and MIT's Gerd Ceder and Yet-Ming Chiang. They were all stars in their own right. The Argonne leadership—Mark Peters and

Isaacs himself—had to be handled as well, as did the industry players. Could Chamberlain do all that? Madia had his doubts.

Isaacs thought that Chamberlain possessed the raw skills, and told Madia so. The situation was not as dire as he claimed. But there *was* something to Madia's message: if the proposal was going to come together, Chamberlain did need to up his game. He was exceptionally smart; when he put his mind to something, he could dig extremely deep into the science. He was technologically and entrepreneurially savvy—in the IP department, no one was better at putting together a deal. But while those were admirable traits, they did not add up to exceptional leadership—not when it came to something of the scale of the Hub proposal. In the Battery Department, Chamberlain had done well at managing a couple of stars— Thackeray and Amine. But as Madia said, there were many more heavyweights on the Hub team. From a tactical perspective, Chamberlain had to figure out how to nudge these personalities to work together—to work for *him*. If there was an inflection point on the way to this new and improved Chamberlain, it would have to be now.

Madia asked Isaacs whether he could abandon protocol, under which he was to stand aloof from the proposal writers, and converse face-to-face with Chamberlain. Chamberlain had asked to see him, he said—perhaps it would help.

"Of course," Isaacs said.

Madia appeared at Chamberlain's cubicle. After a couple of niceties from the older man, including the air traffic controller metaphor, Chamberlain began to speak.

"I've cut deals with Intel and Cisco. I've dealt with big important negotiators. We've only interacted a couple of times, but you're the best I've seen," Chamberlain said.

"I don't know about that," Madia said.

"I do," Chamberlain said. "So I'm excited about working with you. But that idea also makes me fearful."

"Why fearful?"

"Because you could end my career at DOE with a single phone call."

"I wouldn't do that."

"I know you wouldn't, but you could," Chamberlain said. "Now I could say to Eric, 'I want to step out of this and go back to managing the Battery Department.' But that's not me. I'm going to see this through. This competition could make my career. I could move on to the next, then the next. But I want you to know that fear is there."

"A little fear is a good thing," Madia said.

Chamberlain knew he had turned Madia. Regardless of what the older man thought, it would be he, and not Madia, leading the relationship. He spoke again.

Madia should tone down his manner, Chamberlain said. "There is not much time left. You are a very frank person and extremely direct, which is what we are paying you to do," he said. "But you have to be aware that there are a lot of sensitive people on the team. If these sensitivities are alerted, that might lose us a week, and I can't lose a week."

Madia laughed. He understood Chamberlain's drift and promised to do his best.

The Hub team now had nothing to worry about. Chamberlain in fact did not think Madia was the entire problem—his people were paranoid and hypersensitive. Both sides could do with some relaxing. There *were* things to learn from Madia—many things. Chamberlain had not been merely flattering him. Madia was in fact the best dealmaker he had encountered.

36

War Room

The Hub team moved into a space the size of a regulation basketball court on the first floor of Building 200, across a lawn and a parking lot from Building 205. When you entered, you walked between dark gray cubicles arranged into two rows. All the activity was in the far end of the room, a section separated by a partition and containing a conference table surrounded by soft, black swivel chairs. It seemed intended for important meetings. Chamberlain called it "the War Room." Butcher paper was taped on a wall, each sheet representing part of the proposal. Two columns were sketched onto four additional sheets that hung on the back wall behind a large video screen. In the left was written, "Days till Submission." In the right, "Days till Red Team." The former was fifty-three, the latter twenty-eight.

Chamberlain strode in. He had been delayed by a call with Madia.

"What visionary ideas are we seeking?" he said.

They needed a core rationale for their claim to the Hub. Chamberlain said the question had to be answered in the first five pages of the proposal. The Hub would be won or lost there. If the initial pages were loved, a reviewer would read the next twenty-seven subsections simply to justify the positive impression. But if they failed to do that—if they could not get the first five pages right—all would be in vain.

Gallagher asked, "Are we writing to the FOA or are we reading between the lines?"

"What do you mean?" Chamberlain said.

Gallagher was responding to Madia's admonition that Argonne had overemphasized basic research. His observation stumbled on the team's central dynamic: for the six weeks they had been meeting, Chamberlain's guys had plumbed a single question—what did the Department of Energy *really want?* There was the FOA of course—the department's own instructions—but surely there was more to it. Chamberlain led this suspicious line of discussion. He would introduce his thoughts with preludes such as, "Our intelligence is telling us that . . ." or "Our guy in the DOE is telling us. . . ." No one knew the identity of these mysterious sources, but Chamberlain's tone suggested he had the real scoop: the FOA said applied, but it meant research.

Madia had told the group, "Believe the FOA." In other words, Chamberlain might have his sources, but they were wrong.

So at this moment Gallagher stared intently and spoke slowly, a loose hand gesturing. "Well, until now we've been anticipating what we think is between the lines. So are we doing that or writing what the FOA actually says?"

Chamberlain turned to the rest of the group. "In industry," he said, "I always sought to find out what the customer wanted and fulfill it. In this proposal, I had resisted that—until this review." He trailed off but the meaning was clear: the old operating principle was dead. They would stop second-guessing.

He flicked a slide on the screen. Its headline was "Goals." Then it said, "Rapid evolution. Operating prototypes of currently non-existing technology."

Chamberlain said, "All the work, including research, is aimed at producing functioning prototypes. I mean all work. We aren't going to produce quantum dots for the sake of making quantum dots. But if we can argue that making quantum dots will allow us down the road to create a functioning prototype, we will."

Brad Ullrick, an Argonne lawyer, sat near Chamberlain. Neither had actually competed to sit on the proposal team. The pair had swooped in and claimed seats in the Hub. No one else *sought* the positions they now occupied. The two men had unusual skills for a national laboratory—Chamberlain an entrepreneur, Ullrick a legal heavyweight with a background in patent law, both sharp and enterprising. Ullrick was recently divorced. He and his former wife lived a couple of blocks from each other and shared custody of their two young daughters. He was athletic and personable, with a warm smile, and wanted to date. But the bruises of the incompatible marriage were conspicuous in his diffidence. Ullrick seemed to emerge most fully in Chamberlain's presence. "Jeff's my best friend," he said.

A year and a half earlier, Ullrick had suggested to Chamberlain that he—Chamberlain—was destined to be an assistant lab director, a direct deputy to Isaacs. At the time, it seemed an outlandish idea. It was true that Chamberlain had made the unusual jump from the intellectual property unit to Battery Department head. But senior management of the lab as a whole? Now Chamberlain was directing what could be the most consequential project at Argonne in memory and Ullrick's call seemed less farfetched.

Ullrick thought it would be healthier if Argonne had more people like them. Perhaps, but it would open up a new set of demands because the lab would then have to figure out how to hold on to these very different character types. If you looked at Chamberlain's record, he had not stayed with his jobs for long—a few years. No one could know if he would stick around this time. But meanwhile Ullrick planned to ride up with him.

Coghill attempted to orchestrate the team's methodology. Once, she wanted the group to sit in threes and pass edited pages from one person to the next, providing all the chance to mark up each section. It could take awhile but was extremely effective, she said, resulting in a well-scrubbed document. Instead, a dozen or so team members worked on their sections alone.

One day, Isaacs walked in behind a delivery boy with a stack of pizzas. He told the team that what they were doing was important and made small talk. The effort helped—the Hub team seemed bucked up by Isaacs's personal interest in the project. Isaacs also noticed a change in Chamberlain, a new decisiveness. It was how he spoke to his people, including to Isaacs himself. Respectful, but clear. "You'll do this for me," he would say. Chamberlain had made the necessary transformation.

Someone asked Chamberlain how the proposal was looking. "Crappy," he said. Madia had told him to expect to finish just three quarters of it by the time of the Red Team review, and that seemed about right. Was he feeling the pressure?

"Yep."

Madia's follow-on review came and went. Chamberlain thought he was again "brutal" but also "very helpful" in suggesting specific editing changes. Someone had finally crafted a snappy vision. It was called "five-five-five." The idea was to achieve five times greater energy density than current state-of-the-art batteries at one fifth the cost, accomplished in the five-year program duration.

George Crabtree, a superconductivity pioneer and member of the National Academy of Sciences, would run the Hub if Argonne won. Crabtree had been added to the Hub team relatively late. The original choice—a university scientist—backed out, which turned out to be a good thing because, while a strong researcher, he was uninspiring and lacked influence beyond the labs. The battery stars would not have followed him and he would have little influence in Washington, which after all would select the winner. That had led Isaacs to Crabtree, a well-liked former Chemistry Division head and one of Argonne's main ambassadors to the Department of Energy. Crabtree had a gentle way about him and a gift for articulating complex ideas; senior Department of Energy supervisors frequently invited him to contribute to important energy studies and lead public events. What he lacked was any knowledge of batteries. But Isaacs thought he would quickly get up to speed.

Crabtree had the scientific stature that the Argonne team needed and so was the right guy. The Red Team agreed.

Thackeray and Amine dismissed five-five-five as fantasy. But the Red Team liked the sound of that, too.

The initial five pages were still not ready. The Red Team "trashed us" on that section, Chamberlain said. It was crucial that Argonne explain how it would meet its goals. Chamberlain said, "We have to rewrite the whole thing."

Some of the other criticism could be blamed on Chamberlain's leadership flaws. His inclination to seek consensus was slowing down the work; it was time to take the proposal out of the group and put it into the hands of two or three decisive team members. Generally speaking, Chamberlain said, he needed to "listen to my instincts better." In one case, the Red Team had suggested he use a chart rather than words to make a point. Within the proposal team a few days earlier, Chamberlain had suggested precisely that approach but had relented against resistance. "I listened to everybody else. Damn it," he said. Chamberlain said he did not plan to become a dictator but would manage differently. He said, "I am becoming more and more aggressive and confident. Let me put it that way."

He was experiencing mood swings. Immediately after the review, he concluded that the effort did not stand a chance. It simply could not be brought up to par by the May 31 deadline. Four hours later, he said, "It is looking pretty good. We are going to win."

Battery guys traveled to Washington, D.C., every May for the Department of Energy's week-long Annual Merit Review, when the recipients of its approximately $90 million in annual research spending stood before judges to defend their work. This year it was held at a Marriott in Virginia, just across the Potomac River from D.C.

Voltage fade dominated the first day. Peter Faguy, the Department of Energy official, said fade "has become front and center" given the reward to the electric-car industry should NMC 2.0 deliver the promised performance. Amine, Thackeray, and Chris

Johnson rose one after another to propose separate solutions to
the problem. In the back rows, Gallagher worried about the Hub
proposal, due in just over two weeks. So did Jack Vaughey, playing
Angry Birds on his smart phone. Both had been elevated to the
core writing group, leaving behind their routine duties, and had
been working sixty hours a week on the proposal.

That evening, Gallagher and Croy, Thackeray's protégé, joined
Don Hillebrand, the automotive group director, for a dinner of
Mexican food. Hillebrand pointed out that politics played a considerable role in large funding decisions and that no one should be
surprised if Oak Ridge—a recognized heavyweight—bested Argonne for no reason other than good political or bureaucratic connections. He went on a bit longer, when Gallagher said, "I don't
appreciate what you are saying." Gallagher said he had been pour-
ing his heart into the project, and here Hillebrand was making
light of it. Whatever the case, Hillebrand should shut up. Hillebrand apologized and said he hadn't meant that at all. The dinner
ended a quarter hour later.

Outside, Gallagher texted Chamberlain, who said Hillebrand
was a "dick" and had "always been a dick." Then he texted Tony
Burrell, who said Hillebrand was an "asshole" and not to worry
about it.

Standing near another door, Hillebrand said, "Either those guys
are really nervous or he is an asshole," meaning Gallagher. "And I
suspect it is more of the latter."

With ten days to go, Chamberlain, Mark Peters, and Holly Coghill
huddled in Argonne's War Room. They had raised $52 million in
commitments for outside funding on top of the $120 million Hub
award. Their thousand-page proposal would include personal let-
ters from Rahm Emanuel, Chicago's mayor, and California governor Jerry Brown, along with the governors of Illinois and
Michigan, twelve United States senators, and a considerable list of

congressmen. They did not expect to produce a perfect document. Nor would they write to the bitter end. One of the final decisions was a superstitious one—they would submit the proposal ahead of time. They would take no risks with the deadline.

On May 25—six days before the due date—Chamberlain hit the send key.

37

Getting to a Deal

Two years earlier, in 2010, GM had found that Kumar's version of NMC missed a couple of important specs, specifically with regard to a metric called DC resistance. This malady arose when you sought to travel the last twenty or so miles on a one-hundred- or two-hundred-mile battery. The material would put up fierce resistance—the cathode would object to taking in the last of the lithium from the anode. It wanted you instead to plug in the car. The sensation for the motorist was that you suddenly lost power—the vehicle would become exceedingly sluggish. GM said that Kumar had to significantly reduce the DC resistance in his cathode before it could be installed in a commercial car.

But that did not dampen GM's enthusiasm for the cathode. The advantages should Kumar succeed were too tantalizing to ignore: his new electrode, more than any that GM had studied, seemed to be on course for a large improvement over the best commercial cells. If he resolved a few hang-ups, his cathode would allow the carmaker to significantly reduce the price of its electrified vehicles and increase the distance they could travel on a single charge. GM engineers continued to counsel caution—given the risks if anything went wrong or if Envia simply fell short of expectations, the carmaker was better off relying on long-standing supply relationships. But Kumar's ARPA-E announcement—the disclosure that he had broken through existing performance barriers and created a 400-watt-hour-per-kilogram battery—had tipped the discussion.

The Envia announcement had been somewhat exaggerated—the start-up suggested that it was achieving 400 watt-hours per kilogram of energy density for hundreds of cycles.[1] In fact, it maintained that milestone for just three cycles before the energy plunged. But this seemed to be overlooked—anyway, it would improve. GM management wanted the Envia battery and ordered its negotiators to close a licensing deal.

About this time, Kumar e-mailed a slide deck to Damon Frisch, GM's designated intermediary with Envia. It was a blockbuster follow-up to the ARPA-E Summit. Envia, one slide said, had achieved "precise control" of the anode's silicon-carbon nanostructure. It was starting to scale up production to two-kilogram bulk batches.

The previous summer, GM's Mark Mathias had told the Argonne battery guys that Kumar's NMC 2.0 "does not work" as yet. About nine months later, Mathias said he remained on the fence—Envia's product *might* yet get all the way there or, conversely, could "become the sort of thing that sends people back to the fundamental R&D labs." Lauckner—who held a decisive vote—said he was impressed with Kumar's progress. The silicon-carbon anode was "probably by now close to six hundred cycles, which is pretty respectable," he said. "You will run into a lot of battery company start-ups that will be happy if they can get it to one hundred cycles." His negotiators continued to hammer out a deal with Kapadia.

Word somehow reached the Department of Energy that Envia was a foreign acquisition target. Honda and Toyota in fact were hounding Kumar for custom NMC 2.0 cathodes and more energetic batteries, and Samsung's Sun-Ho Kang had told him that he wanted the juiced-up electrode as fast as possible. The DOE sent an alarmed message to Chamberlain: stop Envia from migrating abroad. Chamberlain was furious. Envia itself was looking for a cash exit. It was not the target of a hostile takeover. And if Envia wished to leave the United States, that was "none of our business." "I just look at it this way," he said. "We are a bunch of scientists working for bread. Why are we thinking about industrial policy?"

Chamberlain was either naïve or pretending. If the Department of Energy sought to hold on to Envia, a federal grant recipient with a significant apparent breakthrough, it was acting no differently from Chamberlain himself. Wasn't he forever speaking of saving America? Wasn't that what the Hub was all about? The Hub *was* industrial policy. Sitting in his office, Chamberlain conceded that this was true.

Argonne and Envia were not the sole U.S. actors in the race. While they fought for prominence, Elon Musk, the South African chairman of Tesla Motors, became the popular face of electric cars in the country. A lithe, distant, and tall man with furry patches around the perimeter of his face, Musk had earned a fortune by cofounding and selling PayPal. Now, with his exquisitely designed Tesla, he had made electrics cool. To power them, he had in a way endorsed the ExxonMobil forecast: he had snubbed the race for a battery breakthrough and staked his ground on what was available off the shelf. Musk's bet was that a pure engineering play—a sizzling concentration of high-tech luxury on wheels—could win the market before anyone created a super battery, perhaps long before.

It was a brave, clever, and altogether unpredictable maneuver. Musk, a doyen of Silicon Valley with a bachelor's degree in physics, was thumbing his nose at scientists: his senior team seemed old school, with a former Toyota executive in charge of manufacturing and a Mazda man as chief designer; only JB Straubel, his chief technical officer, was a standard product of Silicon Valley, with a master's degree from Stanford and a string of technology jobs. And though they invented no new battery materials, their cars were unlike anyone else's. They were propelled by "18650s," cylindrical nickel-cobalt-aluminum batteries with the same general appearance as AAs made for cameras, only larger. The batteries were acquired from Panasonic, some eight thousand in each car. They added about 1,300 pounds of weight to the vehicles but, mounted in the floorboard, they contributed great stability. Using the 18650s meant exchanging development risk and cost for engineering risk: though Musk's team didn't have to struggle with physics and invent the better battery, they did have to design a battery *pack* that delivered increasing efficiencies until a model could be sold for roughly $30,000, Musk's goal for a mass-market electric by the end of the decade. As of now, Teslas sold for double and triple that sum, depending on the model.

For the very reason that his cars did not require a laboratory breakthrough, Musk captured the rapt attention of big incumbents. Toyota, Mercedes-Benz, and Daimler variously bought shares of Tesla and signed deals to acquire its power trains and batteries. It was powerful validation.

Musk said he was not worried by NMC 2.0. Kumar had said he was speaking to Tesla, but Musk remarked, "There are a lot of claims made by battery people." His team had selected nickel-cobalt-aluminum based on price—it was the cheapest calculated by kilowatt-hour, he said. If he were leading the development of the Chevy Volt, Musk said, he would immediately discard the NMC and switch to his own material.

The Argonne guys disputed the wisdom of Musk's choice—Gallagher said that nickel-cobalt-aluminum, while having impressive energy density, was also among the most volatile of the main lithium-ion chemistries. It easily caught fire. Musk was courting trouble by putting it in the Tesla.

At ARPA-E, Kumar had unveiled a battery that produced 400 watt-hours per kilogram of energy density. But he would need to seriously improve its cycling performance to make it usable in an electric car—the battery would need to be capable of being charged and discharged 1,000 times. One major hurdle was voltage fade. Thus far, like Musk, he was relying on an engineering solution to stabilize the fade—he was keeping voltage under 4.5 so as not to trigger full-blown fade. You could use the resulting NMC in a working battery. But performance would be much better if it could be reliably activated at 4.6 or more volts—if Envia could deliver NMC 2.0. The start-up could spark a titanic jump if it truly got to the bottom of voltage fade.

Kumar wanted to act fast. It was time to shift from engineering solutions to the deep science. He needed someone familiar with state-of-the-art nanoscale beam-line instruments, technology available only at the national labs. Only such an expert could conceivably deliver those results.

Kumar invited Jason Croy to pay a return visit to Newark.

38

"So What Is Wrong with Me?"

Although his nameplate remained on the corner office in the Battery Department, Chamberlain had his family photographs and other personal belongings removed to the War Room. He would now work full-time on the Hub, so a replacement would have to be found for him in the department. The shift coincided with a perceptible change of sentiment among the battery guys. In prior decades—as long as anyone remembered, anyway—every head of the Battery Department had come from outside. Since Elton Cairns in the 1960s, no one had ever been elevated to the top post from within. In fact, said Jack Vaughey, there had been no promotion of any kind—apart from postdoctoral assistants hired as full-time staff—in some three decades. Now Tony Burrell, a recruit from Los Alamos National Laboratory, had been hired as interim successor to Chamberlain, and several of the battery guys were unhappy. "So what is wrong with *me*?" was how one of them put it.

Neither Thackeray nor Amine coveted the wholly administrative post, but they, too, bristled at Burrell's elevation. Unlike Chamberlain, who confined himself to activities like licensing their inventions, Burrell pushed into the science. Both battery geniuses griped when Burrell—not himself a battery guy—took charge of the department's lithium-air research effort. Amine expected to run lithium-air, especially since he had attracted considerable funding from Dow Chemical for part of the work. Thackeray said that you simply did not appoint a non–battery guy to a supervisory role over something as serious as lithium-air. Amine was still more piqued when Burrell took command of the voltage fade team as well. Sitting straight across the table from Burrell in a big departmental meeting, Amine said the job should go to Chris Johnson. He said Johnson's knowledge of NMC surpassed almost anyone else's. Burrell was like a potted plant—Amine did not so much as glance at him. The rebuke not only of Burrell but also of upper management's prerogative to select group leaders fit Amine's personal style. Its bluntness stood out.

Burrell's multiple appointments irritated Amine, but the main rub was stature and respect. In a series of private meetings, Amine asked Chamberlain to consider his achievements: at the age of forty-nine, he was a 709, the second most senior rank in the national labs. He should be a 710, what was known as a "distinguished fellow," he said. Each large unit at Argonne was apportioned one 710 slot or occasionally two—there were two in Building 205—but no more. There were a total of just ten 710s or so for the lab as a whole. This pinnacle was the crowning achievement of a long career, the recipients typically closing in on retirement, in their late fifties or early sixties, such as Thackeray, who himself was a 710. But Amine said he should not have to wait—he had won numerous awards, published "tons of papers," been granted some 120 patents, and was recognized internationally. He was as good as Thackeray. "You know, listen. I've been very patient," he said. "You can talk to the upper guys. I'd like it to come from you so they know that you are supporting me and playing your role." The unspoken threat hung in the air—if Chamberlain would not argue his case, he would go over his head and "make things happen." Amine had a record of doing so.

A few years earlier, Amine had approached Chamberlain's predecessor, Gary Henriksen. "Gary, it's been four years since I've been a 708," he had said; it is time for promotion. But Henriksen had refused to recommend him; it was not out of disrespect for his work. It was that for two years Henriksen had been pushing the promotion of another senior lab researcher, a man older than Amine who Henriksen felt deserved the bump-up first. Stacked up against Amine's application, which is how the lab management might handle two such simultaneous recommendations, the other man would stand "no chance" of promotion; Amine would overshadow him. The fair thing was to allow the other researcher to go first, then look at Amine's case. Amine insisted.

"What an ingrate," Henriksen had replied, and rejected Amine again.

Amine went to the director of the chemical division. "You promote me," he said, "or I'll resign. You make your decision. I need to know within a week." The next day, the phone rang. "Get your papers in order. We will move them forward." Both men were promoted.

Not long after the blowup over Burrell, Amine was dining on sushi when his cell phone began to vibrate. "Oh, hi, Jeff," he said, excusing himself.

"You got your promotion," Chamberlain said. Chamberlain was not going to let the situation boil over as before. He had presented Amine's unusual request up the ranks and won approval.

Amine returned to the table, a broad smile on his face. He said, "Jeff is a great guy. He's a guy who recognizes—I think we are very fortunate to have him."

Later, Emilio Bunel, the division chief and Chamberlain's direct boss, summoned a Battery Department meeting. He had an announcement: Jack Vaughey was herewith promoted to group leader. If the stars were publicly grumbling, Bunel and Chamberlain wondered who was next. The department employed some sixty researchers but had the same management structure—a director and Amine and Thackeray as group leaders—as when it numbered only a half dozen. Contentment did not seem possible when people remained in exactly the same jobs for their entire career. Vaughey would now be in charge of his own group of battery researchers.

Once the shock wore off, the move seemed to lift the pall. Amine's promotion was not yet public knowledge, but the announcement about Vaughey was enough to create a titter in the department. Some toted up a private list of others deserving a bump, including themselves. None complained not to be first because Vaughey, a fifteen-year department veteran, was so diffident and affable.

Vaughey seemed grateful but said his life was unchanged. He did not receive a raise because Argonne salaries had been frozen along with those of all federal employees for a couple of years. He would carry out precisely the same work, which was the continued attempt to make a pure lithium metal anode that did not catch fire. If he succeeded, it would be a colossal achievement, one larger than a solution to voltage fade. It would bring more recognition than any conceivable promotion. Still, Vaughey said such a move—someone's elevation—was overdue.

Jason was unmoved by Vaughey's promotion. More than halfway through a three-year postdoctoral assistantship, he had a personally pressing matter. If the lab did not intend to promote him as well—with a staff position—he had to start looking outside. He was almost a decade older than many of the postdocs and had his two children to think about.

Thackeray said that Croy merited promotion but that there was currently no opening; even if there were, he would have to compete for it. The reality was that Argonne, like most employers, could act fast if faced with losing someone valuable. But the risk had to be explicit—there had to be an actual competing offer. Croy said that while that might be true, he would not enter into a job search lightly. If he received a job offer and liked it, he would accept and leave Argonne. "I wouldn't pull a gun and not be willing to use it," he said. "I wouldn't waste my time or theirs." Thackeray said he understood.

Croy e-mailed Kumar, which led to an appointment with Envia. The interview took place just a few months after Croy, Thackeray, and Kumar had exchanged views on voltage fade. His work was already familiar to the Envia team. But as they watched Croy flip through his slides, they saw that this was new stuff. Croy and Gallagher had recently obtained fresh clues from the beam line as to the nature of voltage fade. It was precisely the sort of progress that Kumar hoped to tackle at Envia. When they subsequently spoke one on one, however, Kumar did not question Croy on science. He rather spoke generally about the profession—about finding someone who would fit well in Envia. Croy could see that Kumar cared "about the company and its image. They like what they are doing and believe in it."

Closing out the day, Croy met with Kapadia. The CEO said that everyone Croy had met at Envia had liked him. "We are going to give you an offer," Kapadia said. "If you want the job, you should ask for whatever will make you happy."

Croy was delighted. Here was an enterprise clearly on the move, managed by talented and motivated scientists with faith in themselves. Kumar's substance had especially made an impression. He still favored a position at Argonne but was eager to hear Envia's offer.

Argonne reacted fast. Chamberlain—no longer leading the Battery Department but still in overall charge of energy storage at the lab—said that Croy was equivalent to Gallagher. Both were "stars." "If we lost Jason, it would be hard on us, because his potential is great, not only in science, but to lead," Chamberlain said. "There is always a small percentage who can lead and also have the intellect. He has both."

Notwithstanding his usual modesty, Croy thought the same. In the subsequent days, Thackeray asked where Croy saw himself in future years. "In Dr. Isaacs's seat," Croy said.

Chamberlain said that one thing was clear for both Argonne and Envia: whoever had Croy would possess "the guy who is closest to the needed breakthrough" on voltage fade.

A little over a month later, a two-page letter arrived from Kumar by e-mail. As a personal touch, the Envia founder left a message on Croy's office phone. He hoped that Croy would join Envia as a "senior scientist." Kumar threw in thirty thousand share options.

In an IPO, they could be worth $500,000 or even $1 million, Kumar said, enough for a serious down payment on a decent house even in the expensive East Bay. Croy declined to disclose the financial details to Argonne but said the package was "very attractive."

Chamberlain fashioned a counteroffer. He knew he could not match Kumar on salary dollar for dollar, but he could come close when you considered the respective costs of living. Housing was far cheaper south of Chicago. Neither could he offer anything resembling stock options. But he knew that a big factor for Croy was the proximity to his and his wife's families in Indiana. Lindsey Croy didn't want to move to California. Chamberlain hoped that and the higher salary would tip Croy's decision.

The Argonne bureaucracy was slow to approve the proposal. A week after Kumar's offer was in hand, Croy said, "Still waiting to hear from Argonne if you can believe that. No job, no Hub, no money. Only thing we have is voltage fade, as constant as the stars!"

He had two days left to reply to Kumar.

Chamberlain and Croy spoke by phone. Chamberlain said the younger man should make the best decision for his own circumstances. Envia's offer was extremely attractive, perhaps superior financially to the one to come any day from Argonne. But, in the spirit of friendly advice, he wanted to make sure that Croy understood, in case he didn't already, the reality of the Bay Area. "There are enormous risks going to Envia," he said. "Look at Solyndra. It could be shuttered in two years. Silicon Valley eats people up." He went on:

> At this point—pre-IPO—it is intoxicating. You can believe the hype because you want to. They dangle options in front of you. But there is no way to predict if Envia is a mirage. Even if you knew all the data. I am not talking Envia down—I think they have a great shot at it. But if you are outside of it entirely, you just don't know what the truth is.

Chamberlain was sincere. But, notwithstanding his doubts about Envia, Thackeray wondered if Chamberlain had gone too far. "Can you imagine if we convince Jason to stay and five years from now he could have been worth $5 million?" he said.

At last, Argonne's human resources unit e-mailed a counteroffer. Croy had another day to weigh the two.

The next morning, an e-mail from Croy arrived in Chamberlain's in-box. "I just wanted to write and officially accept," he said. Envia's offer was rich, but Chamberlain's instincts were strong. The Croys had lived away long enough; Lindsey Croy refused to do so again.

Kumar understood. Deciding between the two offers had "torn Jason apart," Kumar said. Yet he regretted not having managed to persuade the Argonne man. Croy did as well. "I thought it would be a good thing" to go to Envia, he said.

39

"Throw Out the Old Paradigm"

I n July 2012, the Argonne team traveled to Washington for the Hub orals. This was the toughest part of the competition. The reviewers, consisting of battery specialists and scientists from unrelated fields, would be demanding and could be brutal.

The Argonne guys were worried by a celebrity team from Oak Ridge Lab and the University of Texas at Austin. It was led by Ray Orbach, director of the university's Energy Institute. Orbach had an inside track. He was a former undersecretary for science at the Department of Energy, an accomplished theoretical and experimental physicist with deep connections in the scientific community. Orbach had recruited a powerful roster, crowned by perhaps the most dangerous personality of all—John Goodenough, the eighty-nine-year-old lithium-ion battery pioneer. Chamberlain could say that Argonne's proposal featured the most illustrious battery team on Earth, but it was a frivolous claim when competing with Goodenough, the father of modern batteries.

Goodenough was competitive and gravely serious about the Hub. The breakthroughs to be targeted were "critical for the social fabric," Goodenough said. "If you are going to go beyond seven billion people on the Earth, you need to feed them all."

On the favorable side from Argonne's point of view was that Orbach, having signed the field's living legend, had gone on to antagonize him. Goodenough was threatening to withdraw from the team. His gripe was that Orbach's proposal managers were treating him like an artifact rather than the active battery man that he continued to be. "They would like me to be associated [but] in a way that they run everything," he said. "You know, young Turks want to impose old-man honor and not power." He laughed but was genuinely unhappy. Goodenough said, "The fact of the matter is I don't want my name used improperly."

Oak Ridge's internal dissent aside, the Argonne team couldn't count on their rivals' blunders. They had to focus on their performance in the orals.

Eight-member teams were permitted in the orals room. The Argonne team would consist of five presenters from across the country—MIT's Yet-Ming Chiang; Argonne's Nenad Markovic, an experimental chemist from Serbia; Berkeley's Kristin Persson, a materials specialist; and of course Crabtree and Chamberlain. That left three spaces open. Argonne's managers put much thought into who else would lend the proposal the most gravitas; what chords had they not yet struck that might appeal to the reviewers? It was decided that Dow Chemical, an Argonne partner, would receive one of the open slots to demonstrate industry's endorsement of the proposal. Khal Amine would take the second as a recognized battery superstar. That left one slot. Chamberlain nominated Gallagher, who would be evidence of the new generation coming up. But at the last minute, Isaacs stepped in to take the slot for himself. He wanted to show that the entire lab was behind the effort. Gallagher could travel with the team to Washington as a utility player on call outside the room. At least two decades younger than the other team members, Gallagher was thrilled merely to be present.

Then there was the proposal itself. As the weeks had gone on, Gallagher's techno-economic model kept moving closer to the front of the document and now was a growing factor in Argonne's orals presentation. Just the day before, the group was rehearsing at the lab in front of Isaacs and his deputy Peter Littlewood when Chamberlain reached the middle of the slide deck. The two Bell veterans leapt at Chamberlain's description of the Nelson-Gallagher model.

"This has to be right at the beginning," one said. "This is what we have to talk about."

With each such elevation, Gallagher obtained an effective promotion. As it now stood, if Argonne won, he would be catapulted to a management slot just under the senior men.

Gallagher himself considered the model an "incredibly useful tool." But he had suppressed his opinion in early Hub discussions, conditioned by the gibes he usually heard around the lab, scorn that the model was "second-rate stuff that anybody who has a bachelor's degree could do." The other battery guys shunned the model, which was why Gallagher was "amazed" now as Isaacs and Littlewood talked it up as an ingenious way to build prototypes on the cheap: you would accelerate innovation by quickly informing inventors what would and would not work at factory scale.

Littlewood's own excitement went back again to Bell, where he recalled an effort to build a prototype that had inadvertently led to the discovery of the "fractional quantum Hall effect," a principle of physics that won another Nobel Prize. Littlewood endorsed the Hub's intention to emphasize prototyping as a first principle.

To Gallagher, senior management's attitude clarified the Hub's core mission, which was "to throw out the old paradigm and say, 'If you could start from scratch, how would you organize things?'" The writing team was unified in this arguably radical conclusion—that to rapidly launch the age of electric cars, you needed an entirely *new system* of research. You had to ignore virtually all the battery work currently under way and conceive both a new way to invent and new concepts of storage, a frame of mind that Goodenough himself would recognize. "Most of us believe that you would eventually get there with today's model," Gallagher said. "But it would be very inefficient. We want to try to do it in the near future instead of the distant future." If something out there was going to inadvertently create the big leap, it would have already done so. The only way to win the race was to start over.

The assertion that nothing currently on the market or in battery labs anywhere in the world was up to the task was a bold one.

Gallagher wasn't even certain that the Hub was the prescription. Unlike Khal Amine, who said that the right team and the right sum of money would produce the battery breakthrough in a decade, Gallagher was "a little more pragmatic. I don't know. The thing is I am really excited about trying."

The Hub might get there, or might not. Gallagher said that battery development as currently practiced at Argonne and everywhere else was a dead end.

The group had rented and catered an extra room at the Germantown, Maryland, Marriott, where they were staying, thirty miles northwest of Washington, D.C. The orals would be held nearby on Wednesday in the Department of Energy's basic sciences division. The team would rehearse in the room today—a Monday—then rest tomorrow.

The team assembled in the morning for this last round of practice. Crabtree went first. The Argonne-led group, he said, had carefully considered the FOA—the announcement of the Hub opportunity—and looked forward to the reviewers' scrutiny. He flashed a chart on the screen. It was a "unifying graphic" conceived by Gallagher that depicted innovation leading from the laboratory to the market. At the nucleus was the techno-economic model. In the Hub, it was relabeled "system analysis," and Crabtree, with the fresh orders from the top, promoted it like a breakthrough innovation in itself. It was "a new paradigm for battery development," he said, an example of "science-based rational design" through which "we can predict what we will get."

Crabtree said the current methodology, where scientists published articles and engineers focused on performance, resulted in isolation between groups working on the anode, cathode, and electrolyte. "They operate on a hunt-and-try approach. So if something works well, 'Let's do it.' If it doesn't work, we will drop it without trying to understand why," he said. The result was battery performance improvement of about 5 percent per year, "which is actually quite impressive, but *not* fast enough to achieve the goals."

To meet the Argonne team's transformational goals, there had to be a new way of doing things, Crabtree said. He unveiled five-five-five—five times greater energy density at one fifth the cost, carried out in five years. "So let's look at how that system would work," he said.

As a given battery design was studied, researchers would reach a point at which the Hub managers would step in and caucus: did the results merit a go-ahead? If the answer was yes, the next stage would involve a "translational development team." Crabtree called attention to a slide. "These colors represent the representatives of every part of the Energy Storage Hub," he said. This team would now move the promising battery concept along to "more decision points." If the concept continued to meet the benchmarks, it could then graduate to industrial scale-up.

The team was not starting with a blank slate of new battery theories. The proposal had three potentially transformational concepts, each of which, in the opinion of the Argonne team, stood a reasonable chance of working. To plumb these concepts, the team had marshaled "game-changing tools," Crabtree said. They included "the Electrochemical Discovery Lab," to be equipped with every major measurement tool required to evaluate a battery. These tools were sensitive—you had to be trained in how to use them. But once you were, the tools had the power—along with techno-economic modeling—to carve months and years off the time required to separate winning from losing chemistries.

Some concepts would not make it. The team would determine that they were unsuitable for development. "In fact, we expect to have many of these," Crabtree said. Some of the rejects would be nonetheless mined for useful knowledge—lessons learned that would be driven back into the interrogatory process through a specially created "Science Library of Knowledge."

"We've assembled a team consisting of five national laboratories, five universities, and five industry partners," Crabtree said. "One team operating under one management with one mission."

Chamberlain presented next. He spoke of the "temperature plot," which Argonne believed would be crucial to the Hub's success. It was part of techno-economic modeling. The idea was to track every material under study on a chart. Color and line length would indicate the material's readiness for translation into a prototype. The longer and greener the line, the readier it was in terms of performance, cost, energy, power, safety, and life. Importantly, the data were comprehensive. By cost, Chamberlain did not mean just electrode preparation, coating, and so on, but shipping and receiving the product—the total cost of developing a battery pack. A high-voltage system, for instance, could almost certainly operate in relatively small factories, have a cost advantage, and thus be evident in a long, green line.

The reviewers needed to contemplate another important question as they compared Argonne with its Hub competitors, Chamberlain said. That was "how quickly can you get going on day one?" Not only did the Argonne team have "the skills and the people set to do it. We already have the equipment," he said. All the members had signed a legally binding agreement governing intellectual property. "We don't have this fear" of losing out if an idea is swept up by another team member. Chamberlain said, "We can openly collaborate because we already have a plan." A single entity—Argonne—would control all licensing. The Hub would be "a one-stop shop for industry," he said, and "hit the ground running."

On Wednesday morning, July 18, the Argonne team filed into a battered room in the DOE's basic sciences division headquarters. Inside sat the reviewers, empty seats for the Argonne guys, and an old projector to display slides.

The projector didn't work. Chamberlain summoned Gallagher, who, standing outside the room, had a spare. Then the room went dark—somehow the electricity had gone down. The battery guys stood still. Long minutes passed.

Finally, the lights returned.

Chamberlain said, "I think that was eight minutes. We'll get that back, right?"

You can have five, a reviewer said.

The Argonne team silently cropped out lines.

After the orals were over, the group gathered for coffee. The reviewers, particularly Jeff Dahn, the competing inventor of NMC, had seized on the five-five-five claim. "There is no way you can do this," Dahn said. Other reviewers picked up the theme. Chamberlain had faced precisely the same doubts at Argonne; before an uprising could break out among the reviewers, he stood and said, "I learned in industry and academia that if you set goals with people or for people, what people do is they naturally try to achieve that goal. So we set it." He nodded to the reviewers. "It is going to be extraordinarily difficult to achieve this goal. And we understand it is aspirational. But if we make it halfway or two thirds the way there, we will shift the market. And I don't mean it will move things a little bit and create businesses and jobs. It will shift the market."

Chamberlain hoped his point was understood—that if they were ultrarealistic and set a goal for a 50 percent improvement over lithium-ion, the researchers would aim precisely for 50 percent. But if they targeted a fivefold improvement and got twofold, "that is a monumental achievement." A couple of the battery guys were unhappy with Dahn, but Crabtree thought his questions were fair—tough, but fair.

The flight back to Chicago was in a couple of hours. At the airport, Isaacs said that no one could have commanded the floor as Crabtree had. When he entered the room, every member of the review committee knew who he was. "We couldn't have done it without you," Isaacs said. Chamberlain said that everyone "really did a good job."

Someone asked, "Was it a slam dunk?"

"No, not at all," Isaacs said. His voice was tight.

"We are going to wait and see," Crabtree said.

It was pointed out that prior to the orals, Thackeray had forecast an Argonne victory. Chamberlain laughed.

"It is not slam-dunk," Isaacs said. "There will be other strong competitors and you can't tell. These committees can do anything that they want to do. And DOE has its own mind. It is going to be a real tough decision." No one in particular could know what mastery Oak Ridge would put on display.

Not a slam dunk. But they felt damn confident.

40

The Waiting

Three weeks after the orals, a curt message circulated among Argonne's senior managers including Crabtree and Chamberlain: assemble for a conference call with Isaacs.

The call started an hour later.

"Are you sitting down?" Isaacs asked. "Well, here is a piece of news. We won the Hub." The long struggle was over. The Hub was Argonne's.

Isaacs barely let the news sink in. They had to transform the proposal into working plans for execution as soon as the Department of Energy made the decision public. That could be any day. They needed to be ready. Meanwhile, they were sworn to secrecy. No one else, even other members of the proposal team, could know. Isaacs himself would pass word according to protocol—after the Department of Energy announcement.

The rest of the team was anxious. Amine said the decision was certain to come in a few days or surely early the following week.

Amine meanwhile was wheeling and dealing. His immediate goal was to get NMC 2.0 into the iPhone and other Apple electronics. He had met with the company's battery guys in Cupertino, California. Apple products relied on workhorse lithium-cobalt-oxide batteries. They were unquestionably reliable. Still, they were not the best that Apple could do, he said. Their capacity was 140 milliampere-hours per gram, of which you could access only half because of deterioration over time. The average iPhone could run several hours while using just the telephone but much fewer if you were running Google maps. NMC 2.0, conversely, gave you 190 milliampere-hours per gram, more than 30 percent better capacity. It operated at higher voltage, which meant even greater energy. NMC 2.0 still required work, including coating and additives to protect the surface of the electrodes. But Amine said switching would be a smart strategic move for Apple. The issue was not only performance. So many electric vehicles were going on sale that they would eventually place a serious strain on the global supply of cobalt. The price of the metal was likely to soar. That was not a huge matter for the car-making industry because cobalt made up just 30 percent of the cathode in car batteries in favor at the moment. But the current Apple cathode was 100 percent cobalt oxide.

Apple dispatched a team to Argonne to examine Amine's suggestion more deeply. "This is a big deal if Apple licenses. It is huge," Amine said. "They told us they will need one billion batteries within the next five years. Amazing." He laughed.

Two months passed. No Hub decision was announced. The senior team began to wonder whether someone's mind had changed. Even those in the know said they could no longer be certain of what Isaacs had confided. Maybe they had not won after all. Or perhaps the decision had been overturned. Who knew, given Washington politics and the presidential race?

"What time is it in Washington now—two forty-five?" Mark Peters asked, standing outdoors one day with Thackeray. "It won't happen today. It'll be next week."

As the weeks wore on without an announcement, apprehension deepened. Despite assurance from up high, the core group worried that somehow another team had managed in the silence to reverse the decision. All they could do was wait. "If we don't get the Hub, it will be a major disaster," Amine said one day.

In October, A123 declared bankruptcy. It had been on the verge of selling an 80 percent stake to a Chinese company—Wanxiang—when it decided instead to file for Chapter 11 protection. The Argonne team fretted that the bankruptcy could be a political dealbreaker—A123 had received $135 million in federal stimulus money to fund a Michigan battery factory. On the very day it filed the bankruptcy papers, it received a check for $946,830 from the Department of Energy. The Argonne team acted forcefully and told the Department of Energy that A123 had withdrawn from the Hub team. But no one could know whether that would be regarded as sufficient inoculation against political attack should an Argonne victory be announced.

Crabtree heard from someone in the Department of Energy whom he would not identify that the announcement would probably wait until after the presidential election on November 6. It was easy to infer that no one wanted to risk a scandal that might jeopardize Obama's reelection.

Steven Chu was due in Chicago in a couple of weeks. Gallagher hoped that someone would muster the courage to ask who won. Mostly though, Gallagher said he felt naïve. He and the rest of the team had nourished grand expectations for the proposal. They had assumed that outsiders would buy into their idea. Political reality appeared to be stepping in the way. They had never contemplated that risk. "I think it is referred to as 'drinking the Kool-Aid,'" Gallagher said.

Obama won easily. Yet there was still no word. "We are going to hear something before Thanksgiving," Chamberlain said. But there was really no knowing. The battery community was flowing with rumors. Chamberlain said that if he plotted the various purported announcement dates on a graph, the distribution would probably be clotted around the current week—in the middle of November. But some said the announcement would not come for another four months. "I just don't know what to believe," he said.

At last, official word arrived. A conference call was arranged with Argonne management—the lab had won. The announcement would be made November 30 at the University of Chicago.

That morning, Chamberlain and Crabtree pulled battery components one by one from bags they toted into the lobby of the Gleacher Center, the modern main building of the University of Chicago Booth School of Business. They were props to explain the Hub's work to the invited dignitaries. Isaacs waited farther inside, in a staging area for the speakers. Chu arrived before long, eager and alert. Chicago mayor Rahm Emanuel was next. Their bonhomie, forged in Obama administration cabinet meetings, was instantly apparent. Isaacs tried to break in by telling Emanuel he was delighted to deliver the Hub for the city.

"You didn't deliver it," Emanuel shot back. "*I* did."

Emanuel did not appear to be joking. He seemed certain that his White House connections and not the Argonne proposal tipped the decision. Who knew—perhaps Emanuel *was* crucial in the choice. It did not matter at this point.

When the ceremony was over, Peters, Isaacs's deputy, pulled Chamberlain aside. "This was the best moment of my career," he said. He paused for a long moment, then pointed a finger at Chamberlain. "And it is because of you. Thank you." Chamberlain felt terrific. He said, "Right back at you, man."

Isaacs said an albatross had been lifted. If Argonne had lost, "it would have been everyone's neck and guess whose fault it would have been?" he said laughing. "There were times when I sweated a lot, but this is what this is—a competition." Isaacs would make sure that whatever lessons Chamberlain had learned would be codified so Argonne could win future big competitions with less effort—and angst.

Five days later, Isaacs stood before the whole of the Battery Department and the larger Hub team. They were gathered at the lab guesthouse for wine and hors d'oeuvres. Isaacs thanked each of the main actors by name. He told the story of how, after Argonne had lost the three big prior competitions, Bill Madia had said,

"What the hell is wrong here?" Madia had not necessarily discovered what went wrong, Isaacs said, "but he did figure out how to do it right." The group shared a big laugh over the story of Emanuel's claiming credit for himself.

Just beforehand, Thackeray had passed Isaacs in the corridor. The director had looked at the battery genius and flashed five-five-five on each hand, one after the other, before going on and out the door. Thackeray laughed.

41

Deal

About five hours after the ceremony at the University of Chicago, Atul Kapadia and Sujeet Kumar hosted the Envia staff at a Palo Alto restaurant for the annual holiday party. The pair had been waiting for days for General Motors to e-mail a contract, only to go home without it. The talking had gone on so long and with such uncertainty that neither man had even told the scientists that they seemed to be on the verge of their first licensing agreement. Even if they felt more confident, they could not have said anything, since such news could affect GM's share price. But word had leaked around the Newark lab anyway. An edginess hung over the lunch.

Three months earlier, GM CEO Dan Akerson had himself described Envia's exploits in a closed meeting of the carmaker's employees: in just a couple of years, he said, Envia had created a breakthrough battery that was certain to power an electric car a hundred miles on a single charge. And "we've got better than a fifty-fifty chance to develop a car that will go *two hundred* miles on a charge," Akerson said. That longer range could be achievable by around 2016, which, he said, "would be a game changer."

Akerson believed GM and Envia were partners in this potentially world-beating push to the future. "These little companies come out of nowhere and they surprise you," he said. He did not reveal the tense licensing negotiations under way. But, back at Argonne, Thackeray suspected something afoot when he read of

Akerson's remarks, which had leaked. He said, "I believe that Envia may announce a deal soon—perhaps an order?"

A month later, Hari Iyer, Kapadia's deputy for the GM account, had begun to circulate proposed contract language among the senior Envia executives: "supplier manufactures lithium-rich layered-layered composite cathode materials and silicon-carbon anode materials for use in high-performance rechargeable electric battery cells."

"I am okay with it," Kumar responded to the September 27 e-mail.

The draft contract went on to be quite specific: Envia was to provide a working 350-watt-hour-per-kilogram battery that could endure one thousand charge-discharge cycles. The requirement was not 400 watt-hours per kilogram—what you achieved in a cell could never be matched when you scaled up to the battery pack that actually went into a car. But it remained a tremendous challenge, since Kumar still had to overcome the silicon expansion on the anode side, along with the old issue of DC resistance in the cathode. The deadline was October 2013. After that, adjustments could be made to optimize the performance until August 15, 2014. But that was a full-stop deadline—Kumar could make no changes to the battery after the latter date. This point was critical to GM because once the battery was known to be in place, all the other deadlines could follow, ending with the 2016 launch of an electric car with two hundred miles of range on a single charge. As with the more reasonable demand for a 350-watt-hour-per-kilogram battery, GM was aiming not at the three-hundred-mile car of which Kumar had spoken in Orlando.

"I am fine with the deliverable and time line," Kumar wrote Iyer on October 25.

While the lawyers continued to talk, GM was becoming nervous. In addition to the two-hundred-mile electric into which the 350-watt-hour-per-kilogram material would be placed, GM had to place Envia's other advances into production for the 2015 model year Volt. It was understandable why Envia insisted on keeping its signature ARPA-E material under wraps until the contract was signed—it was too valuable to let out. The start-up was almost obliged to allow no one direct access to the material. But GM wanted Kumar and Kapadia to brief LG, GM's battery cell manufacturer, on the lower-level NMC cathode, the one that would go into the 2015 Volt.

It was still an unusual request: suppliers did not ordinarily open up their proprietary inventions absent a sealed contract. Much could go wrong with a deal even in advanced negotiations. But the Envia men said they had no problem with the favor, given the goodwill between the two companies. Envia disclosed the secrets behind the plug-in hybrid version of its NMC, a configuration that reduced the cost of the cathode 30 percent, to $230 or $240 per kilowatt-hour. That could allow GM to immediately reduce the price of the Volt by $3,000 or $4,000.

GM and Envia continued to haggle over how and when to announce the deal publicly. One idea was for the main players to assemble for a ceremony either outside Detroit or in Newark. That suited Kumar and Kapadia, who were eager to get the news out because the publicity could induce Honda and possibly Toyota to sign follow-on licensing agreements. GM, however, worried about the politics—Republicans still relentlessly labeled the Volt the "Obamamobile" and stigmatized electrics generally as publicly subsidized playthings of the liberal elite. The carmaker said it would delay public disclosure of the deal until after the presidential election.

As the election passed, the lawyers were still swapping sentences without closure, but a couple of weeks later, they seemed to have arrived at an acceptable version. Crucially for Envia, it would receive $2 million a quarter, adding up to $8 million a year, for at least four years. It was sufficient to pay all of its bills—its entire burn rate. On top of that would be royalties once the cars began to be manufactured. In all, the deal would run for eighteen years. Depending how many cars were sold, the deal could be worth hundreds of millions of dollars. The board was delighted. It awarded new stock options to twenty-five employees, about 45 percent of them to the three-man business team led by Kapadia, and promoted Hari Iyer to vice president.

So much time had passed, however, that there seemed no way to schedule a formal signing event. Given the stringent deadline, every day counted. So the document would be signed by GM without public fanfare at its Warren, Michigan, office, then e-mailed to Kapadia for his signature.

The Envia men waited.

Kapadia's cell rang as he drove back from the holiday party. It was General Motors: senior management had finally signed the documents. They were on their way by e-mail. Kapadia turned off the phone.

A regular office meeting had been scheduled back at Envia. Kapadia said it would be delayed a few moments. He tried not to let on.

In the conference room, some papers lay on the table before Kapadia. It was the company's first licensing deal, he said—with GM. "I always knew that Sujeet would build a special technology. He is a special guy," Kapadia said. Kumar and his team had a lot more work to do if the 350-watt-hour-per-kilogram battery was to become reality and enable electric cars. But "the only company that has any chance of getting to that promised land is Envia," he said. Only Envia could change the world. "Just to let you know, this is not my achievement. This is your achievement. And I am signing on behalf of you. I want to sign here in front of you guys." Kapadia bent over and initialed the papers.

The room erupted. The three dozen scientists and staff cat-called and screamed. They jumped up and down. To some of those present, the tumult sounded like two hundred people.

Now Kumar spoke directly to his scientists. "The business guys have delivered. I have never seen a business team deliver like this. Now it is our time. The onus is on us," he said. "There are no excuses. Get this done. Put the chemistry in the car. Whatever small issues are remaining need to be solved in the coming year."

Days later, Kumar said he was "very happy" with the contract. The months ahead would be "pretty tough," he said. "I would say the time line is aggressive." But after all the make-work deals in which Envia was producing tiny batches of material for coin cells and enduring excruciating scrutiny, he had finally proven himself. The company was earning serious revenue. It was no longer a start-up on the come. Envia had arrived.

42

The News from Envia

T he day after the deal closed, Kumar received an e-mail from Damon Frisch, his GM program officer. Frisch would arrive in Newark two days later—on Monday. He would have his entire team with him. They would start work immediately.

The idea was to begin with the launch date—2016—and work backward. Over the following two days, Frisch, Kumar, and their teams would go through every major step of the laboratory process until they had a meticulous schedule of milestones. In just twenty months, this time line would deliver the central nervous system for the electric that Dan Akerson had promised his employees.

Such painstaking planning was vital from GM's perspective because most cars did not earn large margins for their manufacturers. If GM was to profitably make electrics, there was very little room for error. Kumar said that to deliver the required performance, he would have to start with a better understanding of his materials, using equipment he did not currently possess. He would require access to either synchrotron radiation or neutron diffraction capability, either of which would permit him to peer deeply into his battery. Only the national labs had this capacity—the beam lines at Argonne, Lawrence Berkeley, and Oak Ridge. Kumar would need to obtain beam time and to do so absent the usual three- and six-month-long waiting times.

It would help if GM finally announced the deal. That would supply Kumar the credibility to enter into a conversation with the labs about beam time—to request and be granted favors from them. He thought the labs should waive the usual fees given the industrial and corporate stakes. "I feel this should be a national mission," Kumar said. The national labs should treat the NMC 2.0 and the silicon-carbon anode as priorities. "Why shouldn't they work on something that has an American company and is to America's advantage?" he said. "Why shouldn't they analyze my material as opposed to publishing many, many papers that are fantastic scientifically but don't help American industry?"

The GM and Envia teams endorsed the schedule. Now Kumar embarked on the dizzying sequence of milestones. He assumed that disclosure of the deal would follow more or less immediately in order to provide him some negotiating traction with the labs.

Kapadia figured the deal would be made public by early January at the latest. He reckoned on a resurrection of the previously discussed public ceremony; seeing as how the two-hundred-mile car would validate Obama's goals set out three years before, the White House itself might participate in the news conference. That would help his own efforts. As Kumar advanced on the GM contract, he moved forward on the plan to allow Envia investors to cash out. There no longer seemed to be time to launch an IPO. It again seemed much more sensible to find a buyer. Potential American purchasers remained cautious, so Goldman Sachs took the acquisition proposal to Asia, sending queries to, among other suitors, Japan's Asahi Kasei and South Korea's LG and Samsung. These companies had to feel that time was of the essence—that Envia could be swept up any moment by a rival. The GM deal would help create that impression.

But GM again said the announcement had to wait. There was an electronics show the first week of December, which was not a good time for the disclosure. Then the Detroit auto show—also not ideal. After that, Envia, LG, and GM were to be occupied in a brainstorming session in Detroit. That would push back the announcement to the third week of January. Then there was the second Obama inauguration. Which made the latest plan to wait until the next ARPA-E Summit, in February. There would be something poetic about that date given the announcement at the last Summit.

GM was not announcing the deal but neither was it moving back its time line. The deadlines to deliver the working battery remained etched in stone. "I feel pressured," Kumar said. He had to find and hire a structural cathode scientist, someone like Jason Croy who was familiar with the beam lines. Since A123 had declared bankruptcy, Kumar figured that he could raid it for at least some of the ten or fifteen scientists he required. He would place an ad and start making calls.

The ARPA-E Summit passed without an announcement. But toward the end of February, Hari Iyer received an urgent e-mail from General Motors: it had finished initial tests on Envia's 400-watt-hour-per-kilogram battery. So far, they had failed to replicate the results announced at ARPA-E. No remedy seemed to work. Could Envia explain why?

Increasingly concerned queries piled up during routine conference calls and meetings with the GM team over the next week. The cell was not reproducing the ARPA-E results. It was supposed to be delivering 400 watt-hours per kilogram, and it wasn't—not by a long shot. Iyer asked Kumar, who shrugged off his questions. Iyer approached members of the scientific team. Finally he went back to Kapadia. Iyer said, "Perhaps what we told the world and GM isn't what it is."

Kumar's technical team in fact had an inkling of what GM was finding. Two weeks after the ARPA-E Summit the previous year, they had sent another cell to Crane for evaluation. When the result was returned to Envia, it verified the 400-watt-hour-per-kilogram claim—but only on the second cycle. After that, the energy density plummeted. By the fifth cycle, energy density was down to 302. On the hundredth, it was at 267. By the two hundredth, it had dipped to 249, and on the 342nd cycle—the last listed—energy density was at 232. The cell had lost 42 percent of its energy and showed no sign of stabilizing.[1] But the GM team seemed not to achieve even two cycles of the 400-watt-hour-per-kilogram performance.

As part of his explanation for the battery's problems, a staff scientist had told Iyer something unnerving. Envia, he said, was *not* the inventor of the 400-watt-hour-per-kilogram battery touted at ARPA-E. Not of the entire technology, anyway. The NMC cathode was legit—Envia, as it had told everyone, had transformed the Argonne electrode into an optimized, top-of-the-line product. But not the silicon-carbon anode—the flaunted component that lifted the performance of the battery as a whole was not Envia's invention.

A year before, cofounder Mike Sinkula had told Kapadia that the anode contained material from a Japanese supplier. Kapadia had dropped the subject after it was waved away as unimportant by Purnesh Seegopaul, the materials scientist on the board of directors. But this fresh claim, heard as GM was raising a fuss, assumed a different meaning. It could not be ignored.

Iyer again confronted Kumar, who said that what he heard was wrong—the anode *was* Envia's. He was shipping out the anode for treatments such as chemical vapor deposition, the first of five discrete production processes to which the electrode was subjected. The anode was going to Japan for that purpose. It was all routine stuff.

The explanation made sense—much went into the creation of an advanced electrode. If an outside supplier painted a GM car, for instance, no one would challenge the assertion that it was in fact still a GM car. So it was with the anode. Kumar said Envia did the work that mattered.

Kumar said that the most pressing problem was not the anode, but how much voltage GM insisted on applying to the battery. The greater the voltage, the more lithium began to shuttle between the electrodes and the farther a car could go on a single charge. That was the idea behind activating the NMC at 4.5 volts and thus transforming it into NMC 2.0. But such higher voltages also stressed the material and made the Argonne battery fail—it was what caused voltage fade. Kumar wanted to stay well away from the stress limit. For ARPA-E, he had pushed the voltage as high as he dared—to 4.3 volts—and at that level managed to cycle the battery 450 times; to meet GM's specs, he would have to more than double that to 1,000 cycles, but Kumar was confident he would succeed.

Only, after the deal was signed, when the GM and Envia engineers sat down to map out the work to come, the carmaker surprised him by insisting on the application of *4.4 volts*. While that tenth of a volt may have seemed only a tweak, its impact was magnified on the atomic scale. At that state of charge, atoms began to move around at an accelerated pace, the cathode expanded and contracted with the shuttling of the lithium, and ultimately the material could crack. You began to lose electrical contact between the particles that make up the cathode. Kumar began to achieve at best three hundred cycles, much further from the thousand that he needed. Now his team had not to double the battery's longevity, but to triple it.

Kapadia remained suspicious. On March 7, 2013, he briefed the Envia board about their primary customer's unhappiness with the signature product and the doubts about the anode. In an e-mail, he said, "GM has observed a significant and very large disparity between the data obtained from Arpa-E cells and proof of concept cells based on [the] 400 wh/kg technology we sent them for testing." Until the performance questions were resolved, there should be no further discussion of any news releases or contract ceremonies, Kapadia said.

A week later, Kumar held a heated meeting with Iyer. Kumar disparaged Kapadia for putting his concerns in writing, which made them legally "discoverable"—any potential buyer of Envia would have to be provided such internal correspondence during the due diligence process. It was a bizarre outburst. Iyer said, "If there were no misrepresentations, you should not worry about Atul putting these issues in writing."

Kumar broke down and told the truth: the anode—the one used in the ARPA-E battery and shipped a year later to GM—hadn't been serviced in Japan. It had been bought there. What Iyer

heard was true. The anode was made by a company called Shin-
Etsu.

But creating an anode was only the first step, Kumar continued
to say. This being silicon, you had to carry out many additional
procedures in the lab in order to make the anode usable. It required treatments, coatings, and nano-processing. That was why
he regarded the anode as Envia's. It was analogous to the Argonne
cathode. Because Envia had optimized it, Kumar could legiti-
mately call it proprietary.

Kumar asked Iyer to prevail upon Kapadia to withdraw his al-
legations about the anode. If Kapadia would do so, Kumar would
apologize and make peace with the CEO.

It was a bewildering admission, and to Kapadia an altogether
baffling justification when Iyer reported it to him. For two years,
Envia had trumpeted its work on silicon. It had done so to potential customers, in particular GM, to peers at conferences, and to
the world at large. The ARPA-E announcement was only the most-
noticed version. Now one had to wonder what was true. Kumar
had it wrong, Kapadia said. If Envia said that the anode material
was proprietary, as it had done in its license with GM, then it had
to have created the anode. If it was someone else's, Envia was
duty-bound to explicitly disclose that fact. The anode, plain and
simple, was not Envia's IP. Absent the anode, Envia was no better
than the crowd. Panasonic, Samsung, and Argonne itself all were
delivering energy density of 215 or 220 watt-hours per kilogram
using a standard graphite anode. Kumar's NMC cathode combined with a graphite anode—unactivated so as not to trigger the
voltage fade—was *not* superior, not by more than a few percentage
points anyway. Kapadia himself was at fault for championing the
battery for so long despite the countervailing data and the complaints. But now he was going the other way: Envia, he said, should
come clean in order to try to salvage the relationship with GM.

The Envia executives and board skirmished between and among
themselves on what to do. Kapadia and Iyer insisted that the board
allow them to go to GM and fess up. But the directors ordered a

brave face and continued work. To Kapadia, Seegopaul seemed less concerned about the anode's origins than that its discovery could upset the chances for his venture fund to cash out.

At the end of March, the two companies—GM and Envia— assembled for their first quarterly review. GM's contingent was led by Matthus Joshua, head of purchasing for the company's hybrid program, and Larry Nitz, its head of hybrid engineering. The GM men were blunt. In addition to the shortcomings of the ARPA-E material, Kumar had missed his very first milestone—the provision of one hundred kilograms of NMC cathode powder for the 2015 Volt. They asked Kumar what was going on.

Kumar responded that the problem with the cathode powder was temporary. There had been "a cross-contamination issue" and the Volt commitment would be met. The GM men were "dismissive [and] mocked Kumar," Kapadia said later. They demanded that he reproduce the ARPA-E results.[2] Kumar had lost credibility. This was not the friendly repartee of prior company-to-company talks.

As for Kumar's story about chemical vapor deposition, the GM men said that if that was so, they wanted to see a full accounting of materials used in the anode.

Kumar conceded the true story as he had earlier with Iyer. With that, Kumar's painstakingly presented picture fell apart. The vaunted silicon-anode performance was at best the flawed, undisclosed work of a Japanese American composite.

What was the answer for the GM men? One could be angry at what seemed a possible case of flimflammery. But from their perspective, how would that enable the vehicles on which a high-profile part of GM's future—not to mention perhaps their own jobs—depended? If Kumar could step up and deliver the ARPA-E performance, GM could sort out the licensing of the Shin-Etsu anode, even if they'd come to it in roundabout fashion.

Joshua and Nitz gave Kumar another quarter. It was no longer important who had invented the silicon-carbon anode. The numbers were the crucial issue. If Kumar could produce the metrics boasted of at the Summit, all could be forgiven.

One thing was certain—none of this, nor the deal itself, was to go public. Nitz said, "The next cell—build or bust" for Envia.

Kapadia's aggressive outing of the anode seriously strained his relationships within the company. He and Kumar now barely spoke. The board instructed him to restrict his activities to selling the company and resolving a lawsuit by Kumar's former employer, NanoeXa. The previous month, NanoeXa CEO Michael Pak accused Kumar in a court claim of stealing intellectual property in order to create Envia's cathode. Kapadia said it was a nuisance suit. But even a nuisance suit could complicate any sale of the company.

Kumar continued to dwell on the potential for a buyout and hence a big payday for investors and employees. July, he said, would be "a fork in the road" when Envia would have to make its decision on its new owner.

Goldman Sachs returned with gloomy news—as of now, no one was prepared to acquire Envia. LG and Samsung had declined almost instantly. Asahi Kasei, the most likely buyer, had also said no. One problem was that the word seemed to be out regarding Envia's underperforming cells. Another was the general malaise hanging over electric cars—even if a big company might previously have been prepared to take a gamble, it was now less inclined given slow vehicle sales.

Kumar had three months of grace time.

43

The Big Man at Argonne

Amid the drama at Envia, President Obama paid a visit to Argonne. When the plans were announced, Lynn Trahey said to Gallagher, "I'm going to find a way to meet him."

"Oh yeah?" Gallagher said. "Can I meet him, too?" He figured Trahey was blustering. But that evening, she e-mailed Matt Howard, Argonne's public relations chief. The subject line was "I volunteer." Trahey went on:

> To spend as much time with Obama as you need me to next week. I know, I know, this is so selfless of me. I think he'd love to meet me! I am 100% available.

Howard replied: "Bribes."

Trahey added rationale for putting her before Obama.

> I'm a female in a land of men. He has daughters so seeing women in STEM will be very important to him. If Argonne is seen as a sausage party we'll be in BIG trouble.
> I'm in the Hub
> I'll blackmail you on bribes
> I'm short so he'll look tall and powerful in pictures.

"All good reasons," Howard replied. Send me a biography and a photo, he went on. But no promises.

Gallagher received an e-mail from Howard, too. Isaacs had endorsed both Trahey's and Gallagher's participation in the Obama visit. They would demonstrate the lab's battery work in a quick two or three minutes as Obama walked through the premises.

Isaacs added Thackeray's name, which was natural. The only problem was that Thackeray was in Barcelona, being feted by the global battery community.

Two months earlier, he had undergone the surgery to remove his spleen. Now fully recovered, he had flown to Spain with Lisa and their three grown daughters, with whom he planned a subsequent vacation around Barcelona and on to London. Thackeray's abdomen was flatter, his smile brighter. Flanked by his daughters, he watched at the Casa Convalescència as, one by one, the world's battery geniuses took the stage to recall his achievements and his friendship. Goodenough, now ninety and caring for his ill wife, could not make it but wrote the frontispiece for the symposium program.

Thackeray took the stage. One thing he wanted to make clear was that this was not a retirement party. "You're not getting rid of me yet," he said. He choked up, talked a little bit more, then, along with Chamberlain, caught a flight back to Chicago. His family would go on without him to London. He wanted to be there for Obama.

The Argonne orchestration machine was in motion. For three days, it subjected Trahey and Gallagher to dry runs before fake Obamas starting with Elizabeth Austin, Isaacs's staff speechwriter. They were coached on where to stand, what to say, and how to say it. The White House advance team showed up and ran Trahey and Gallagher through their lines again.

The morning arrived. Accompanied by Crabtree, Obama bounded across the lab and up to the waiting young scientists. "These [are the] two hotshots here?" Obama asked. "That's them," Crabtree said. Trahey wore her blond hair down. She and Gallagher both dressed in their usual khakis. Each shook the president's hand before Gallagher led off. He pointed to a battery, the Volt's,

which he said "was invented right here at Argonne," then mentioned five-five-five, an account straight out of Crabtree's Hub
introduction. To create the kind of breakthroughs intended in the
Hub, Gallagher said, "you need engineers like me working with
materials scientists and chemists like Lynn Trahey."

"My job is on the discovery side of things," Trahey said, picking
up the narrative. She handed Obama a coin cell while speaking of
chemistry, and herself gripped an atomic-scale model of Li_2MnO_3
made of multicolored plastic beads. They exchanged some chit-
chat. Then Obama was gone.

Minutes later, the president stood before a small, noisy crowd
near the beam line. With a refrain on the global energy race,
Obama unveiled a new long-term fund for energy research. "I want
the next job-creating breakthroughs, whether it is in energy or
nanotechnology or bioengineering . . . to be right here in the
United States of America," he said.

This was about standing behind inventions that might otherwise not happen. Obama said, "Two decades ago, scientists at Argonne led by Mike Thackeray, who's here today. Where is Mike?"

A deer-in-the-headlights freeze crossed Thackeray's face as he
rose slowly up and out of his seat.

"Here he is right here," Obama said, fixing a gaze on Thackeray.
"Mike started work on a rechargeable lithium battery for cars.
Some folks at the time said that the idea wasn't worth the effort.
They said that even if you had the technology, the car would cost
too much. It wouldn't go far enough. But Mike and his team knew
better." So went the story of the NMC and the Volt. Thackeray,
still stone-faced, slid back into his seat.

Obama closed: "We don't stand still, we look forward. We
invent, we build, we turn new ideas into new industries. We
change the way we can live our lives here at home and around
the world. That's how we sent a man to the Moon. That's how we
invented the Internet. When somebody tells us we can't we say,
'Yes we can.'"

It was hokey. A big cheer went up.

Afterward Trahey, cheeks crimson, said, "He stood this close. This close."

"I'm on cloud nine," Gallagher said. "It was an incredible experience."

44

Second Quarter Review

Kumar said his team was "going crazy" attempting to meet the GM specs. The carmaker's aim, he said, was to eliminate its development risk early in the production process so that after August 2014, its sole challenges would be to engineer the remainder of the car and sell it to a tough-minded public. But "it seems we overpromised," Kumar said. "I have never built a team in a rush. It's a recipe for disaster." He thought he could reach 310 or 315 watt-hours per kilogram. That would involve blending the materials from the Volt cathode with those of the materials announced at ARPA-E. If GM wanted to do that, he could have the system ready for the 2016 model. But if the company wanted more, such as the promised 350 watt-hours per kilogram, that "will take another six months or a year," Kumar said. At the higher 4.4 volts, the battery was still cycling just three hundred times. In other words, it would not be ready for the big launch.

He would raise this trade-off with GM when he had the chance. Kumar continued to sound hopeful about Envia's general prospects. The payoff was only a matter of time. He said, "Most likely somebody will buy us out this year."

GM and Envia held their second quarterly review in July. Kumar had again missed the milestones on both the Volt and the two-hundred-mile car batteries. He still could not get the 400-watt-hour-per-kilogram material to perform as advertised. The ARPA-E Summit was a long time back.

The GM men were furious. "The anode material is not Envia's," said Matthus Joshua, the automaker's purchasing executive. Envia had "misrepresented the material." Larry Nitz, the lead engineer, said Envia had earned "a failed grade for this quarter." The meeting ended in acrimony.

Two weeks later, on August 7, a letter from Joshua arrived at Envia. Envia's product claims prior to the contract "appear to have been inaccurate and misleading," he wrote. The anode was represented as proprietary but was actually bought "from a third party."

"Following Envia's admission that it misrepresented the composition, origin and intellectual property content of the 400 wh/kg prototype battery, Envia requested that GM provide additional time for Envia to replicate its 400 wh/kg battery test results. . . ."

GM had provided that time. But, he wrote:

> Envia has failed to move the project forward or replicate the results on a timetable that could conceivably support the vehicle development process. In fact, Envia was unable even to replicate prior reported test results even when utilizing the third party anode that had purportedly been utilized in the Arpa-E test battery.[1]

Joshua meant that even if Kumar could somehow later comply with the promised spec, even if he could later reproduce the ARPA-E result, GM lacked reasonable confidence that it would be ready in time for the 2016 launch. He did not speak of any fallback position for GM. Kumar's private suggestion that they go to a 310- or 315-watt-hour-per-kilogram battery as an intermediate stage went unmentioned in his letter. Given the facts, he wrote instead, the automaker was entitled to terminate the contract. Toward that possibility, it was conducting an internal review—"necessarily re-evaluating its commercial relationship with Envia." GM would notify Envia when the review was complete.

The deal was unraveling. GM said it wanted back the $4 million it had paid out. Envia itself was in turmoil: just eight months after the signatures, Kumar, Kapadia, and leading board members were at each other's throats. The start-up had $2.1 million left in the bank, less than four months of operating expenses.

On August 29, 2013, Kapadia and Iyer caught a red-eye to Detroit. They would try to salvage whatever goodwill remained.

The next day at GM's offices in the Detroit suburb of Warren, they met with a "visibly and justifiably angry" Joshua and Nitz. Kapadia tried to appeal to their sense of decency. Envia had committed mistakes, but the two companies should try "to mitigate the employment and immigration risk on the hardworking scientists at Envia who were likely to be adversely affected" if the start-up collapsed, Kapadia said. The GM men listened. There was no going back, they said—the carmaker was relinquishing the license for the 400-watt-hour-per-kilogram material. But they relented on the more onerous threats. Envia could keep the cash paid. They also signed a legal release—GM would not sue. That was the best they could do.

Kapadia and Iyer flew home, relieved. At least they would not end up in court. Nor would they be forced somehow to scrape together the $4 million.

The following day, the pair reported to the board. Hearing that the threat of a suit was lifted, the directors fired Kapadia, along with Iyer and Rohit Arora, the company's chief financial officer.[2]

Purnesh Seegopaul, the Pangaea Ventures investor, assumed Kapadia's duties. In a meeting with Iyer and Arora, Seegopaul said the business team had done "a poor job negotiating the GM agreement." "The milestones in the agreement were very demanding and . . . no company in the world could achieve them," he said. No one mentioned that nine months before to the day, all three men had been awarded stock options and promotions for a deal well negotiated.

Seegopaul told Kumar to get back to the lab, but under fresh marching orders—he was to make no further public announcements. No one outside GM, Envia, and its board—certainly not the Department of Energy or any reporters—was to know that the doubters about the little Newark start-up had been right all along.

45

Black Box

Kumar blamed the mess on Kapadia, his first investor and the man with whom he had seemed to collaborate so well. During the entire negotiation, he now said, Kapadia and Iyer had run "a black-box operation." Within Envia, they and only they were privy to the details of the deal. The business team had obligated Envia to unrealistic milestones and Kumar to an unattainable time line. The chance for an IPO was likewise squandered.

Kapadia's team included "extra people we never needed," Kumar said, his arms crossed. They were individuals "with a free mind doing nothing" who were responsible for "bad things." Meanwhile, Kapadia had basked in the spotlight. "When something gets achieved, you see the CEO on CNBC. When it goes wrong, you blame the technology guys," Kumar said. "They get the commission, we get the blame. It's always like that."

Thankfully, he said, the board understood what happened and sided with him: "I have a very good board. [Now] people are gone and I can get back to my technology."

And the letter from GM's Joshua and the claims of misrepresentation?

"GM is a large company," Kumar said. "Even if you have one guy who says a bad thing, you can have twenty guys say good things."

Kumar's explanation was all but entirely false. The e-mail record showed that he approved the technological commitments and time line. When he spoke of having "overpromised," the company truly had. But he led that act. All were eager for the deal and guilty of exuberant self-promotion. But Kumar—the company's founder and the discoverer of the NMC in the Argonne catalogue— was the sole executive who knew what was possible. Kapadia and Iyer, in no position to offer expert opinions, checked the dates and material commitments with him. Kumar passed the judgment on which all relied.

Was Kumar a con man? Was he looking to cash out before he was found out? The Argonne guys—all of them skeptics from the time that Kumar began to boast about his big breakthrough— could not decide. Thackeray said Kumar was not a swindler but that, given the temptation to stretch the truth, a battery guy had to "know the difference between right and wrong." Kumar clearly did not. Chamberlain said the truth was obscured by bad blood at the top, but that, if he had to choose who was more at fault, he would not blame Kumar but Kapadia. In his opinion, Kapadia did run a black-box operation, assuming haughty airs while providing no supporting data and demanding that a big American company buy his start-up. Gallagher wouldn't really respond when you asked how he felt. He simply looked sad. "I told you so," his expression said. Envia had fooled many, many people who should have known better. "What is it about humans that makes hype irresistible? Some sort of fundamental desire to dream?" Gallagher said.

At the ARPA-E Summit, Kumar and Kapadia had led a small news conference that included a slide with the first Crane results. Any journalist, if sufficiently acquainted with cycle testing, could study the slide and write the dismal truth about the Envia cathode. But none appeared to be so informed and, absent Kumar or Kapadia calling attention to the plunge in energy density, it went unreported. Kumar's sin of omission crossed the line into explicit deception the same day, when Envia distributed a news release on the 400-watt-hour-per-kilogram breakthrough. This announcement excluded the Crane report entirely and skipped its description of the plummeting energy density. Instead, it published charts that were not part of the Crane evaluation at all. These charts described the anode, which, the illustrations said, continued to work at 91 percent of capacity after three hundred cycles, a very impressive number. The chart was deceptive not only because it displayed data on which Crane did not report, but because it suggested that the cell as a whole was relatively stable.

The trick was the chart's subtle substitution of the metric *capacity*, rather than *energy density*, what Crane measured. Capacity is a measure of how many electrons flow in a single discharge of the battery cell. It can be maintained for numerous cycles even as the cell's energy density is plunging. Capacity was precisely what had deceived Thackeray and Johnson when they were touting their invention of NMC 2.0. The energy of each individual electron—in essence what the voltage measures—was lessening each cycle. This was another way of describing voltage fade. At the three hundredth cycle, the Envia cell's *energy* was at 60 percent of its original level and dropping, nowhere close to the 91 percent of *capacity* that was registered.[1]

For twenty months—from the time of ARPA-E until the fallout with GM resulted in public release of the Crane report—Envia released detailed charts depicting only the capacity. It allowed the impression to persist that it was achieving energy density of 400 watt-hours per kilogram for hundreds of cycles. Again and again, Kumar created the appearance that the cell was fine when the energy density was going haywire. Only when the news of the fallout with GM went on the Internet did he change the company Web site, pull down the charts on capacity, and post the second Crane report. He never released the complete first Crane report—the one on which the ARPA-E announcement was actually based.

Yet that was not the total picture. There was more to Kumar, and when you looked at everything, his behavior did not resemble a pure confidence game. His pursuit of Jason Croy did not square with a picture of a con man. His frustration with the national labs also lacked the appearance of a scam. Kumar genuinely thought he could parlay the Argonne material into a blockbuster new battery and enable the electric-car industry. History is filled with

single-minded entrepreneurs with outsized notions of their prospects. He also was chasing an outsized payday. Along the way, Kumar understood that he was in over his head. But he had made too
many public and private claims. If he just held on, the payday could
come. In the end, as the wreck was approaching, he hoped that
somehow everything would work out. Even if he could not reproduce the 400-watt-hour-per-kilogram super-battery, nor the lesser
350-watt-hour-per-kilogram version, perhaps GM would accept something less.

As for Kapadia, he claimed he had known nothing until the
end, but he clearly had. He knew of the Crane results, which he
himself described to journalists. He was prepared to live with
the deceptive anode: at least twice in the GM crisis, he offered his
resignation but never walked out the door. His team, too, wanted
to stay on despite the faulty performance, the swapped anode, and
the lost GM contract, but were not permitted to.

The board, made up almost entirely of investors and executives,
had personal incentives to stay quiet and hope that Kumar
somehow still managed to win so that they could cash out. Which
is what they did.

As Gallagher indirectly suggested, GM was party to this excessive enthusiasm. Disregarding its engineers, its VC and technology
sides thought their long, painstaking vetting of Envia's coin cells
was sufficient precaution against product risk. But it wasn't. The
carmaker, presumably in possession of the conclusions from the
Crane report, decided that in just twenty months Kumar could
solve electrochemical problems that were vexing some of the best
labs in the world. That was a lot of hope to hang on Envia.

Arun Majumdar, the showy ARPA-E director who had ac-
claimed the start-up to the industry and the world, left the agency
and joined Google. In early 2013, he also became a member of the
Envia board. After it all fell apart, Majumdar said the attempt to
put Envia's work into a commercial vehicle was premature. "Envia
is not ready for prime time in terms of product. That's been the
realization," he said.

As for why Lauckner and the rest of GM elected to stay mum and maintain the appearance of a relationship with Envia after the fallout, that was understandable as well. There was no profit in going public with a fiasco. The news could discredit the Volt and tarnish GM's reputation for developing new technology. Wall Street could pummel the stock. Politically the fallout would also be uncomfortable, as the Republican Party would rub both GM's and the Obama administration's nose in what it would call a waste of millions of taxpayer dollars on useless stabs into clean energy.

Wan Gang was still out there, as were Japan and South Korea. The big stakes remained. So even in a circumstance of misrepresentation, exaggeration, sleight of hand, and general slipperiness, for GM there was a case to be made for sending Kumar back to the lab. He was a scientist. The remote prospect remained that some time in the future, by a miracle, Kumar actually would succeed.

In the weeks after, Atul Kapadia and his two lieutenants asked the Envia board to issue a short press release explaining their departure as a change of corporate strategy and thanking them for their work. If Envia could not see fit to do so, they attached a fifty-two-page civil suit they intended to file in the Alameda County Courthouse against Envia and Kumar personally. It alleged fraud, wrongful termination, and retaliation for their trying to right the false representations about the anode. When the company was slow to respond, the men filed the suit, divulging many of the previous six years of corporate secrets. It was yet another Envia blunder. The mission of keeping the final, sorry year under wraps was blown.

Kapadia's lawsuit borrowed extensively from the legal complaint filed the previous year by NanoeXa, the company where Kumar and his cofounder Mike Sinkula had worked prior to Envia. When examined together, the suits alleged that Envia, while claiming to be marketing proprietary technology, relied almost entirely on IP that it either stole or appropriated without attribution from NanoeXa and Shin-Etsu. In terms of the cathode, the suits, relying on the work of a private forensic investigator, alleged that Kumar secretly downloaded ninety-nine files from the computers at NanoeXa, many of them in a flurry over his last days and hours at the company. Without these files, the suits asserted, there would have been no Envia as it came to be known. Envia flatly denied the NanoeXa assertions. About this time, Kapadia's wife suffered a relapse of cancer, and he and the other two former Envia executives dropped the lawsuit.[2]

46

Back to the Race

After Argonne won the Hub, Chamberlain barely had time to think about celebrating. The Department of Energy insisted on an entirely revised Hub budget—immediately. He thought back over the three years since he first floated the idea for a research effort on the scale of Bell Labs, recalling the history as though telling it for himself as much as anyone else—En-Caesar, the Battery Sematech, and briefing the Obama campaign. He reached back to Cabot Industries, his early experiences in the private sector, and the day he left the start-up world thinking he would coast into Argonne's intellectual property unit for a few relaxing years.

In the ensuing months, Chamberlain would learn that, just as American plans had suffered a setback, Wan Gang had tempered his own aspirations and shifted direction. China's motorists continued to adopt electrics even more slowly than did Americans—some 80 percent of the forty thousand electric vehicles on China's roads in the coming year would be buses and taxis. Wan said he would eliminate a $10,000-a-car subsidy entirely by 2020. The government would instead offer grants to company research labs.

Wu Feng, a senior Chinese official who visited Argonne in late 2013, said that when Wan took up the battery race, he was an electric car expert. China's battery guys had warned him that energy storage was "difficult." But he and the other car specialists weren't absorbing the message. "They imagined that it is easy." Now Wan realized that "it will be a long time to reach the goals." China was still in the race. But Wan had learned that the country would not win quickly.

Some members of the American team felt the same. Peter Faguy, the Department of Energy official who had ordered the crash effort to solve voltage fade, was shutting down the team. Would fade ever be figured out and NMC 2.0 fully enabled? Perhaps, Faguy said. He was not opposed to anyone trying again. But voltage fade had eradicated his own hopes that you could throw money and desire at a problem. "The romantic notion of the line researcher solving it with the light bulb going off is a nonstarter," Faguy said. "These kind of problems are intractable."

Some even said the race was misconceived from the start. Economic and technological hope and coincidence led a lot of nations to chase an illusory prize that then evaporated in their hands. When it was time for the better battery and electric cars, they would arrive and spread quickly, with dividends crossing borders and no single national winner. But that time was not yet with us.

Don Hillebrand disagreed. Extremely capable scientists backed by patriotic governments with concrete and reasonable objectives had made a sincere dash for a better technological path, one that held almost magical powers to resolve some of the era's most intractable economic, political, and environmental problems. When hopes and the stakes for winning are so great, however, you can bring out not only those with potential answers—visionaries—but also "charismatic thieves, swindlers, who are tricking people," Hillebrand said. And, in the case of the battery, when you combined those tendencies with the perseverance of the internal combustion engine, you got a race that "ended in the middle." But Hillebrand predicted that the Hub would invent a blockbuster new battery chemistry. It was a matter of survival. The Hub scientists "know if they don't, they'll be in trouble." Specifically, Chamberlain's "career is very much on the line and *he* knows it."

The War Room was closed. Chamberlain moved into a modern, freshly carpeted suite of offices in Building 200. His new title was Deputy Director of Development and Demonstration for the Hub. That meant that he was responsible for delivering the prototypes promised under the five-five-five criteria. He said he was "highly confident" that he would do so and thus create a new paradigm for American manufacturing. It would be Bell Labs 2.0. He said, "I'm hoping in five or ten years to be touring the country saying, 'This is how it can be done.'"

He watched electrics quietly moving ahead. The Volt for sure was a pioneering vehicle, but Elon Musk had pushed further—he had made electrics indisputably cool. With Tesla, Musk himself was now the most celebrated technologist in Silicon Valley.

Toward the end of 2014, a mini-rivalry erupted: Musk hurtled into a contest with GM to produce the two-hundred-mile electric. He did not *say* he was in competition with GM—in his eyes, that would be demeaning. But after vowing for years to produce a mass-market electric by the end of the decade, he now said he would do so in 2017. He said the car, to be called the Model III, would cost about $35,000. And that fact—the price—put Musk squarely on GM's turf.

Jon Lauckner, still smarting over the Envia debacle, did not say when GM would release its own two-hundred-mile electric, but it would not easily relinquish its market. GM did not desire a direct clash with Musk, given his rare mastery of product style and mar-keting—his pizzazz—but it had one. If he intended to be out with his Model III in 2017, GM would have to have its rival model on sale that year or earlier.

The stakes were clear. The top electrics—the Volt, the Leaf, and Musk's Model S—were selling at a pace of 2,000 to 3,000 vehicles a month each, but motorists were buying about 40,000 of the BMW 3-Series, the entry-level gasoline-driven luxury car that Musk identified as his genuine competition. They would be some-what over the price of the average vehicle, but at that rate of sales, the competing GMs and Teslas would tip electrics into the broad consumer market. They would no longer be niche vehicles. At once, Obama's aim for 1 million electrics on the road would be realized. And that is what Musk said he would do—he alone would sell 500,000 electrics a year by 2020.

LG Chemical, GM's lithium-ion battery supplier, contributed to the drama. Prompted by no one obvious, a senior executive blurted out at an earnings presentation that the company would manufacture a two-hundred-mile battery in 2016.[1] He did not say for whom LG was making the battery, but the disclosure seemed to further telegraph GM's forthcoming battle with Musk.

The pieces were in place. Tesla and GM would achieve their range by re-engineering current-generation battery technology and lightening up their respective vehicles. The Hub, by starting over from scratch, stood a fighting chance of taking batteries the rest of the way.

Chamberlain said the Hub meant peering deep into the physics, with no assumptions as to where the answer lay. Armed with an atomic road map of the chemistry, the United States could really win.

Acknowledgments

When I began to consider a book on batteries, the reception from friends and advisers was all but unanimous: don't do it. It would be a tedious bore. The exceptions were a few hands at Argonne National Laboratory, who seemed to understand the fascination instantly. They are Jeff Chamberlain, Angela Hardin, and Don Hillebrand. I want to single out Chamberlain, without whose support this book would not have happened. The decision to welcome a stranger with a tape recorder into a secure setting with colossally high stakes was not trivial. I would ultimately spend two years in the Battery Department, sometimes hanging around for a few days and once for two months. Chamberlain let go, gave me the run of the department, and persuaded the rest of the lab to relax. They allowed me to be the fly on the wall that I had hoped. I am grateful to Eric Isaacs, Argonne's director, whose permission was required, and David Sandalow, who made a timely call to Isaacs on my behalf that finally set everything in motion. I am also grateful to Sujeet Kumar and Atul Kapadia, who welcomed me to observe Envia's quest to be a successful start-up battery company.

Thanks to New America, which supported this book in an astonishing three-year-long fellowship. New America is a unique institution whose literary and intellectual sensibilities constantly result in outstanding work. Specifically, thanks to Steve Coll, Andres Martinez, Rachel White, and Anne-Marie Slaughter. Superlative thanks to Kevin Delaney, my editor at Quartz, a gentle chief who runs the best publication on the Internet and understood my fixation. In the first fourteen months of Quartz's existence, Kevin accorded me roughly ten months of either part- or full-time book leave. Certainly no other boss would have been so generous. Thanks also to David Bradley, who demonstrates that it is simultaneously possible to be a media colossus and a nice man. I cannot say enough about Jim Levine, the most trustworthy agent and friend in publishing. Jim read and commented on the manuscript three times. Thanks to Rick Kot for embracing the book and to Melanie Tortoroli for her artful and supple advice and editing and ultimately her enthusiasm—she is a terrific friend. Additional thanks to Viking's Ben Petrone and Sarah Janet for their superb work. Susan Glasser commissioned the article for *Foreign Policy* that led to this book and Charlie Homans edited it. Alyson Wright, my research assistant and transcriber, diligently listened to, typed out, and checked hundreds of hours of interviews and offered wise counsel on the content. Noel Greenwood gets an exceptional expression of thanks. Noel edited my first two books and committed to do so this time as well. Throughout, he was in treatment for cancer. Finally, Noel decided enough was enough. I saw him a week before he died and we settled on a road map for *The Powerhouse,* one that I stuck to pretty closely. Thanks, Noel.

I am solely responsible for any errors but I am grateful to Billy Woodford of MIT, who read the full manuscript three times and returned with pages of important scientific and technical corrections, tweaks, and suggestions. Christian Caryl, Konstantin Kakaes, Chris Leonard, Sharon Moshavi, and Chris White read the manuscript and saved me from numerous errors in addition to offering important stylistic suggestions.

A special thanks to those who had the patience to explain (often repeatedly) the science and history of batteries, technology, big geopolitics, and Argonne (with apologies in advance for the names I will no doubt omit for no reason apart from personal shortcomings): Khalil Amine, Jason Croy, Kevin Gallagher, Don Hillebrand, Mike Thackeray, Lynn Trahey, and Brad Ullrick. Also Daniel Abraham, Ralph Brodd, Emilio Bunel, Tony Burrell, Yet-Ming Chiang, George Crabtree, Jeff Dahn, Sun-Ho Kang, Chris Johnson, Peter Littlewood, Paul Nelson, Mark Peters, Venkat Srinivasan, and Jack Vaughey. In addition, Ali Abouimrane, Elizabeth Austin, Mali Balasubramanian, Ilias Belharouak, Martin Bettge, Benjamin Blaiszik, Terry Bray, Rita Brzowski, Zonghai Chen, Holly Coghill, Larry Curtiss, Bill David, Dennis Dees, Carolyn Edmonson, Dan Flores, Sharon Giblin, Gary Henriksen, Matt Howard, Donghan Kim, Greg Krumdick, Jun Lu, Nenad Markovic, Vilas Pol, Yang Ren, Dave Schroeder, Mike Slater, and Cynthia Sullivan.

For kindnesses large and small, thanks to Tom De Waal, Sabine Gallagher, Gideon Lichfield, Sam Patten, Becky Shafer, Faith Smith, Lisa Thackeray, Eric Weiner, and Georgina Wilson.

My most profound gratitude to my family for unconditional support and understanding of absences, silences, and surliness: thanks to Alisha, Ilana and Dolores LeVine, and finally to my best friend and wife, Nurilda Nurlybayeva.

Appendix A

Atul Kapadia e-mailed the following letter to the Envia team on August 30, 2013:

Dear Envia Team:

Thank you very much for all the hard work and the sacrifices each of you made for Envia during the past three years. I greatly enjoyed our interaction but most importantly learnt a lot from each of you.

It was most satisfying for me to watch each of our employees really care about the quality of their work. Everyone worked hard and made enormous sacrifices. When Tom Stephens, GM Vice-Chairman, visited Envia in 2011, he told me in a private conversation that he wished that GM could reproduce Envia's culture—Envia is respected because each employee takes pride in their job and knows that if they did not show up, Envia would suffer. That's high praise indeed. Of course, I did not take any credit for that because the true credit belonged to Sujeet and Herman (for setting up such a work-ethic early-on) and to the team members themselves.

My journey with Envia has been a long one! When I was at Bay Partners, I (along with John Walecka) was the first one to write a check for Envia in 2007 and then re-wrote another check in 2009. When we were approaching insolvency and were using Silicon Valley bank debt money in 2010, I joined the team and Envia closed a highly successful [cash-raising] round of $17 million. Closing those deals with Honda and GM was equally exciting. However, you will all admit that most fun was to compete and win against LG at GM by benchmarking [the] NMC:LMO mixture against our own layered-layered lithium rich cathode.

Some companies have this unique magic potion of employees that as time goes by you see several of the young (and not so young like me) employees become successful via entrepreneurship, via MBAs from prestigious universities or big-shot executives in large companies. I know for a fact that I will hear about many many of you in the years to come.

You all have very exciting individual futures ahead of you. And as long as you keep things simple—ethically and intellectually—nothing will stop you from achieving things that you want to achieve. Surround yourself with people who are tough on you, but also have a strong moral compass.

Some of you have also asked me—what really happened? Easiest way to explain it is when you have strong personalities, there are always disagreements. And disagreements can sometimes be reconciled and sometimes not be reconciled. Add to that 4 months ago, I co-founded a company with a friend that recently got funded. And my co-founder is jealous that I spend all my time thinking about Envia. So as a confluence of several factors, I submitted my resignation to the Board of Directors on August 4 and again yesterday. Today they graciously accepted it.

Just like I was the right CEO when the company had a financing crisis in 2010, our new CEO, Purnesh Seegopaul, is absolutely the right CEO for this moment. He has a Ph.D and has deep experience in the specialty materials world. Despite my several disagreements with Purnesh, in this specialty materials business, I have enormous respect for his judgment. Two years ago, Purnesh gave me some conservative advice on making sure that the business team was not running too fast compared to where Envia was in its technology development cycle. At that point, I disagreed with him. But as time passed, I realized that Purnesh was right. Purnesh is calm, meticulous and methodical—all traits that are vital to running a research company. Make sure you don't let go of him.

Lastly, I received several messages today that I did not address the team in the company meeting. Purnesh graciously offered. But my philosophy is you look forward—not backward. And for each of you the time now is to carve that individual and combined forward path—so all the best!

Regards,

Atul

Appendix B

Envia released the following statement after Kapadia's lawsuit was dropped:

Envia Systems is pleased that the lawsuit brought by Atul Kapadia and two other former employees has been dismissed. Envia is considering its options to address the fact that the baseless lawsuit was filed at all. The company's own investigations and other data confirmed that the lawsuit was meritless, and the plaintiffs and their lawyers have now essentially agreed—by themselves dropping the suit less than two months after it was filed. The plaintiffs are still required to comply with a Preliminary Injunction issued against them in December at Envia's request, which orders them to return company confidential property and data. Envia is moving forward and remains focused on developing breakthrough lithium-ion battery cathode and anode materials for both the automotive and consumer electronics markets.

Notes

2. Why Argonne Let Wan In

1. "Batteries for Electric Cars: Challenges, Opportunities and the Outlook to 2020," Boston Consulting Group, 8.
2. Interview with Peter Harrup, chairman, IDTechEx, July 8, 2010.
3. Ibid.
4. 13 to 15 percent, IHS Global Insight, quoted on Edmunds.com, January 22, 2010; 50 percent, author interview with Ralph Brodd, August 3, 2011. Translating the 2020 percentages into hard figures, we were talking sales of about 7.5 million cars a year. At an average of $30,000 a vehicle, that was a $225-billion-a-year industry, equivalent to the 2012 gross revenue of Toyota, the world's largest carmaker. If accurate, the estimate for 2030 would be more than three times that number.
5. French ecology minister Jean-Louis Boorloo, quoted by Agence France-Press, October 2, 2009.

3. A Good Place to Do Science

1. Arthur Compton, *Atomic Quest* (Oxford, 1956), 144.
2. Jack M. Holl, *Argonne National Laboratory, 1946–96* (University of Illinois Press, 1997), 56.
3. Ibid., 430.

4. "Discouraged Weariness in the Eyes"

1. Lab description and quotes from Robert K. Steunenberg and Leslie Burris, *From Test Tube to Pilot Plant: A 50-Year History of the Chemical Technology Division at Argonne National Laboratory* (Argonne National Laboratory, 2000), 89–160.
2. Ibid.
3. Ibid.
4. Detail and quotes from ibid.
5. C. P. Gilmore, "Electric Autos . . . They're on the Way!," *Popular Science*, December 1966, 76.
6. Stanley Whittingham interview with SUNY Binghamton, October 30, 2000, http://authors.library.caltech.edu/5456/1/hrst.mit.edu/hrs/materials/public/Whittingham_interview.htm.

5. Professor Goodenough

1. John B. Goodenough, *Witness to Grace* (Publish America, 2008).
2. Clare Grey, Barcelona speech, March 10, 2013.

7. Batteries Are a Treacherous World

1. *The Electrician* (London), February 17, 1883, 329.

8. Creating NMC

1. Steunenberg and Burris, *From Test Tube to Pilot Plant*, 470.

11. The New Boss

1. Industry size, Fantasy Sports Ad Network, http://www.fantasysportsadnetwork.com/aboutfantasy.htm.

15. The Start-up

1. Author interview with Michael Pak, December 2, 2013.

16. Out of India (and China and Africa)

1. *San Jose Mercury News*, www.mercurynews.com/business/ci_22094415/asian-workers-now-dominate-silicon-valley-tech-jobs.
2. Organisation for Economic Co-operation and Development (OECD), www.oecd.org/unitedstates/2102002.pdf.
3. *Inside Higher Ed*, www.insidehighered.com/quicktakes/2013/04/03/economic-conditions-key-keeping-foreign-phd-graduates-us.

19. The Car Man

1. Civil suit RG13704405 by Kapadia et al. in Alamada County Superior Court, 12.

29. Orlando

1. Edward L. Morse, "Energy 2020: North America, the New Middle East?" March 20, 2012, Citi notes to clients.

33. ARPA-E

1. Kapadia lawsuit, 16.

34. The Old and the Young

1. *New York Times*, Apr. 13, 2012, www.nytimes.com/2012/04/15/automobiles/how-green-are-electric-cars-depends-on-where-you-plug-in.html?pagewanted=all&_r=0.

37. Getting to a Deal

1. From a public relations . . . Envia "Corporate Presentation" by Kapadia, February 23, 2012. Kapadia presented the fourteen-slide deck to a small group of journalists at ARPA-E on February 27 but it was not released to the general press or released publicly.

42. The News from Envia

1. Private Crane report, June 28, 2012.
2. Kapadia lawsuit, 31.

44. Second Quarter Review

1. Kapadia lawsuit. The author verified the substance of the letter with a source in a position to know its content.
2. See Appendix A.

45. Black Box

1. From Kapadia's "Corporate Presentation," February 23, 2012.
2. See Appendix B.

46. Back to the Race

1. "LG Chem to Supply Batteries for 200-Mile Electric Cars in 2016—CFO," Reuters, July 18, 2014, http://www.reuters.com/article/2014/07/18/lg-chem-batteries-idUSL4N0PT25U20140718.

Index